The Curious Researcher

□ ■ □

A Guide to Writing Research Papers

SECOND REVISED EDITION

BRUCE BALLENGER

Boise State University

ALLYN AND BACON

Boston London Toronto Sydney Tokyo Singapore

For Rebecca,
who reminds me to ask,
Why?

□ ■ □

Vice President, Humanities: Joseph Opiela
Editorial Assistant: Kate Tolini
Marketing Manager: Lisa Kimball
Production Administrator: Rowena Dores
Editorial-Production Service: Susan Freese, Communicáto, Ltd.
Text Design/Electronic Composition: Denise Hoffman
Cover Administrator: Suzanne Harbison
Composition Buyer: Linda Cox
Manufacturing Buyer: Suzanne Lareau

A Viacom Company
160 Gould Street
Needham Heights, MA 02494
Internet: www.abacon.com
America Online: keyword: College Online

Library of Congress Cataloging-in-Publication Data

0–205–29702–1

Source Credits: Figure 1.6 (p. 64) is used with permission of Netscape Communications Corporation. Netscape Communications Corporation has not authorized, sponsored, or endorsed, or approved this publication and is not responsible for its content. Netscape and the Netscape Communications Corporate Logos, are trademarks and trade names of Netscape Communications Corporation. All other product names and/or logos are trademarks of their respective owners.

All other source credits appear in place with the materials they acknowledge.

Printed in the United States of America
10 9 8 7 6 00

Contents

Chapter 1
The First Week 23

Chapter 4 The Fourth Week 155

Contents by Subject

Preface

A few years ago, I had a student who immediately distinguished herself as a gifted writer. Jayne's first essay, titled "The Sterile Cage," was a touching and insightful piece about what it meant to spend a year of her life in a hospital bed, suspended in a stainless steel cage, recovering from a bone disorder. I still copy the essay for my classes, praising the richness of detail and the subtle way a writer's voice can woo a reader. When it was time to assign a research paper, I looked forward to seeing what Jayne would do. She chose a topic in childhood development, and though she said it was going well, I could tell it was loveless labor for her.

Teaching the research paper was the same for me back then. Though I believed in the assignment, convinced that research is a practical academic skill and that research writing is common in other classes, I disliked how cold the classroom became when the subject came up—the sometimes surly silences and sighs—and quickly longed to return to teaching the essay, something for which my students developed some genuine enthusiasm. We would plow through the five weeks it took for students to research and write their papers, all of us anxious for the unit to be over with.

I finally knew I was doing something wrong in my approach to research when I came across Jayne's paper one evening, halfway through the stack of draft research papers. The paper was awful reading. Though she obviously researched the subject vigorously, demonstrating an impressive bibliography, Jayne's paper was unfocused, her analysis lifeless, and her prose wooden. The voice she had found in her essays was "missing in action," as was her usual talent for getting to the heart of a topic. I distinctly remember the conference we had in my office a week later because it was so unpleasant.

What did she think of the paper? I asked. "Not much."

Did she find it unfocused? "Sort of."

Did the paper sound lifeless to her? "Yes."

Jayne glared across the desk at me.

"What do you want from me?" she said. "This is a research paper, damnit. It's *supposed* to be this way."

At that moment, Jayne spoke for the many students I've had over the years who've come to believe that the research paper is a vampirish creation that, at a teacher's request, periodically comes back from the dead to suck the life out of them. Some students, thankfully, don't see it that way, but the Jaynes of the world are not to blame for their cynicism. One student spoke for many when he said, "The definition of *research* is 'tiresome studies on a subject that a person does not like.'"

The assignment clearly carries "baggage," much of it filled with weighty misconceptions about what it means to do research. "Research [in high school] was a certain number of pages with a certain number of quotes on a certain topic," wrote Nancy, a college freshman, when asked about her experiences with research papers. Her comments are typical. "I found it very hard to learn much more than how to link quotes together from these papers. We also had a very strict form for the intro., exactly where the sentence would fall and what it should say."

RESEARCH IS DISCOVERY

In our rush to teach students the formal elements of the research paper, teachers sometimes forget that, despite its many conventions, the success of research writing depends on the same thing that makes most other kinds of writing successful: the writer's interest in what she's writing about. That research can be a process of discovery, a sometimes mad rush down trails that appear unexpectedly, is not well known, even among college students. That researchers sometimes literally hunger to learn more about some aspect of a subject and that the time looking can be fulfilling, even exciting, are less known yet. The reason is simple: Nobody has pointed out that research is really about *curiosity.*

That's the premise of this book: *Research is rooted in curiosity.* This is hardly a radical notion, though strangely enough, many books on writing the college research paper limit their discussion of curiosity to short chapters or sections of chapters, usually titled "Choosing a Topic." Being driven by curiosity involves more than choosing a good topic. Curiosity includes a willingness to be open to what you'll find as you immerse yourself in the research as well as a willingness to change your mind, to let go of preconceived notions. Curiosity is an invitation to be confused and a desire to try to straighten things out, first for yourself and then for the people you're writing for. That's an important distinction. So many students come to the research paper with only one question in mind: What exactly does the teacher want?

CHALLENGING ASSUMPTIONS ABOUT RESEARCH WRITING

Pleasing the teacher is an old game, of course, but it's worse with the research paper because so much of the project seems so prescribed—a certain number of pages, particular citation methods, an outline, an introduction, body, and conclusion. What's worse, many students assume things about how to write the paper that may be unstated—that there is a certain formal voice in all research writing; that all research papers are objective; that the longer the bibliography, the better the grade. With all these rules to worry about, it's no surprise that many students believe their own curiosity about what they're writing is beside the point.

The Curious Researcher says otherwise. I'm convinced that teachers can teach the conventions of the college research paper without students' losing their sense of themselves as masters of their own work. We teachers might begin by reminding ourselves of our own moments of enthusiasm, late at night in the college library, when an unexpected source suddenly cracked open a new door on our research topic and the light poured in, helping us to see the topic in a new way. I have found new enthusiasm for the research paper—in fact, I love teaching it—by renewing my excitement about research as a means of satisfying my own curiosity. It made sense to me to share that enthusiasm with the student readers of this book by relating my own experiences with research. In a break with the usual "I'm the flawless expert" tone of many textbooks, I will also share my own occasional frustrations with the research process. All writers struggle sometimes, including this one.

Despite the obvious differences between the author of this book and its readers, I hope that by sharing the research trails I've followed—dead ends and all—students will see that the research process, like the writing process, is never neat, never as simple as some textbooks imply. Thank God! That's what makes research interesting.

One of the things Jayne was trying to say to me that day in my office was that everything she learned about being a good writer and writing strong prose seemed irrelevant when the research paper assignment came along. Somehow, my students got the message that in the research paper, the writing and the writer don't really matter. In the rigid little world of the research paper, what mattered was getting it right.

This book aims to challenge that assumption by helping students see how integrally the *research* process and the *writing* process are bound. While acknowledging that the research paper does have its own distinct conventions, *The Curious Researcher* promotes the idea that research papers can, in fact, be interesting to write and to read. It accomplishes that, in part, by featuring examples of research writing that are compelling and by encouraging student writers to consider alternatives to the strict forms of the research paper they learned in high school.

TEACHING THE SPIRIT OF INQUIRY

While *The Curious Researcher* features plenty of material on the conventions of research writing—citation methods, approaches to organization, evaluating sources, how to avoid plagiarism, and so on—a major emphasis of the book is introducing students to *the spirit of inquiry.* The habits of mind that good research writers develop is something we can teach that is truly multidisciplinary. That spirit is charged with curiosity, of course—the itch to know and learn and discover. But it also involves the ability to ask researchable questions, the instinct to look in the right places for answers, a willingness to suspend judgment, and an openness to changing one's mind. Embracing the spirit of inquiry must begin with the belief that one *can* be an inquirer, a knower, an active agent in making knowledge.

I think this affective dimension of critical thinking is underrated, especially when it comes to writing research papers. That's why this book promotes the research *essay,* a potentially more subjective, less formal, often more exploratory mode than the formal argumentative research paper. The research essay is, I think, a much better *introduction* to research and research writing and excellent preparation for more conventional academic research because it places the writer in the center of the discourse. As a result, she cannot avoid her role as the main agent of the inquiry nor can she escape the question of her own authority in the conversation about what might be true. When it's a good experience, the writer of the research essay often adopts a new identity as a *knower.*

I am often amazed at what students do with this new freedom. I believe little is lost in not prescribing a formal research paper, particularly in an introductory composition course. As students move on from here to their declared majors, they will learn the scholarly conventions of their disciplines from those best equipped to teach them. In the meantime, students will master valuable library skills and learn many of the technical elements of the research paper, such as citation methods and evaluating sources. But most important, students will discover, often for the first time, what college research is really about: *using the ideas of others to shape ideas of their own.*

WAYS OF USING THIS BOOK

Since procrastination ails many student researchers, this book is uniquely designed to move them through the research process, step by step and week by week, for five weeks, the typical period allotted for the assignment. The structure of the book is flexible, however; students should be encouraged to compress the sequence if their research assignment will take less time or ignore it altogether and use the book to help them solve specific problems as they arise. A "Contents by Subject," new to this edition, makes using the book as a reference much easier.

If you do encourage your students to follow the five-week sequence, I think you'll find that they'll like the way *The Curious Researcher* doesn't deluge them with information, as do so many other research paper texts, but doles it out, week by week, when the information is most needed.

The Introduction, "Rethinking the Research Paper," challenges students to reconceive the research paper assignment. For many of them, this will amount to a "declaration of independence." During "The First Week," students are encouraged to discover topics that they're genuinely curious about and to learn to navigate the reference section of the library through a unique self-teaching exercise. This chapter also includes a wealth of new material on Internet research, including an exercise that gives students a quick tour of its possibilities. In "The Second Week," students focus the topic of the paper and develop a research strategy. In "The Third Week," students learn notetaking techniques, the dangers of plagiarism, and tips on how to conduct a search that challenges them to dig more deeply for information. During "The Fourth Week," students begin writing their drafts;

this chapter also gives tips on integrating sources, structure, voice, and beginnings. In "The Fifth Week," students are guided through the final revision.

In this new edition of *The Curious Researcher,* the details about citation conventions and formats for both the Modern Language Association (MLA) and the American Psychological Association (APA) are in Appendixes A and B, respectively. This organization makes the information easier for students to find and use. New sample papers—one on method acting (in MLA format) and the other arguing for reform of the Endangered Species Act (in APA format)—are included, as well. I like these student essays not simply because they are effectively written. They also offer an interesting contrast between two approaches: one is more "essayistic" and openly subjective, and the other is a more conventional argumentative paper. While the author of the latter paper never uses the first-person singular or autobiographical anecdotes, she still registers a strong writerly presence. In this new edition, I hope to expand the notion of what it means to *get personal* in research writing.

Unlike other textbooks, which relegate exercises to the ends of chapters, *The Curious Researcher* makes them integral to the process of researching and writing the paper. Though techniques such as fastwriting and brainstorming—featured in some of the writing exercises—are now commonplace in many composition classes, they have rarely been applied to research writing and certainly not as extensively as they have been here. Fastwriting is an especially useful tool, not just for prewriting but for open-ended thinking throughout the process of researching and writing the paper. The exercises are also another antidote to procrastination, challenging students to stay involved in the process as well as providing instructors with a number of short assignments throughout the five weeks that will help them monitor students' progress.

OTHER FEATURES OF THIS NEW EDITION

To those of you who know the first edition of *The Curious Researcher,* one difference in this second edition will be evident immediately: Uses of Internet research are now a central feature of the text. It was during a class last year when I realized that some of my students knew much more than I did about online research, and every semester since then, I've noticed growing numbers of students researching on the Internet. It became obvious that if *The Curious Researcher* was to be truly useful to these students, it needed to say a lot not just about the methods of searching the Internet but about the possibilities *and* disadvantages of doing so.

A new section in Chapter 2, "Befriending the Internet," introduces students to Internet research and ends with a self-guided tour of some key sites on the World Wide Web that should be useful to student researchers. This quick tour is perfect for the novice "surfer." Information about the Internet, including the addresses of particular sites, is scattered throughout the book. Because of the explosive growth of information on the Web, a few of these addresses may have changed since this writing. Even so, it still seemed worthwhile to share some of the most promising sites for student researchers. I've also added the latest information on how to cite electronic sources, though as new information and technologies emerge on the Internet, these citation methods will also probably continue to evolve.

Since I wrote the first edition of this book, I've become more and more convinced of the importance of "writing in the middle" of the research process. This is the notetaking stage, often symbolized by a dogged collection of notecards. I no longer believe in notecards, and I've dropped discussion of them completely from Chapter 3. Instead, I've spent more time demonstrating the possibilities of the double-entry journal and the research log. Both of these approaches—and their many variations—strike me as encouraging exactly the kind of often messy, yet thoughtful, writing that notecards discourage. It is here that students take possession of their sources; it is while writing in the middle that students are most likely to initiate *conversations* with the expert voices on their topics that help them successfully negotiate their own authority in their essays.

You will also find some new discussion of how to use the thesis to encourage further inquiry, rather than limit it, along with new material that challenges students to consider not only their assumptions about research but the nature of knowledge. More and more in my classes, I've begun to address epistemelogical questions directly with my students. After all, we're not just teaching them research skills but a whole new way of understanding how knowledge is made and what role they can play in making it.

A number of new exercises have been added to this edition, and I've expanded the treatments of APA and MLA conventions. Moreover, to make this book more useful to instructors who might also be teaching reading-response essays, in addition to research papers, in their courses, Appendix C, "Tips for Researching and Writing Papers on Literary Topics," has been expanded to include a wonderful example of a personal-response essay.

Finally, for instructors who prefer to ignore the week-by-week structure of *The Curious Researcher* and to use it instead as a reference book, this new edition includes a "Contents by Subject." This contents is keyed to the most common problems students confront as they move through the research and writing process.

I'm convinced that the value of any text on research writing, no matter how clever its design, is dependent on the fire ignited by students' topics. Students must be curious researchers. When they take charge of discovering what they want to know, students like Jayne begin to get their voices back, eager to share what they've learned. I believe this book will help.

ACKNOWLEDGMENTS

Teaching writing is often a deeply personal experience. Because students often write from their hearts as well as their minds, I've been privileged to read work that genuinely matters to them. More often than they realize, students teach the teacher, and they have especially tutored me in how to rethink the research paper.

Because students move so quickly in and out of my life, I rarely get the chance to acknowledge their lessons. I'd like to take that opportunity now. Jayne Wynters, whose story begins this book, was a student of mine more than a decade ago, and she deserves much credit for challenging my ideas about the college research paper that lead to *The Curious Researcher.* Other students who either contributed directly to this text or helped me conceive it include Kim Armstrong, Christine Bergquist, Daniel Jaffurs, Heather Dunham, Candace Collins, Jason Pulsifer, Jennifer White, Karoline Ann Fox, Kazuko Kuramoto, André Sears, Christina Kerby, and Carolyn Nelson.

Many of my former colleagues at the University of New Hampshire were instrumental in the genesis of this book. I am especially grateful to three of them. Thomas Newkirk has been my mentor for some years now, and it was his belief that I had something useful to say about the freshman research paper that inspired me to write this text. Robert Connors is not only a preeminent composition scholar but a friend, and his early encouragement and advice were also extremely valuable. And Donald Murray, my first writing teacher, continues to cast his long and welcome shadow on everything I do in the field.

I've never met Rachel Edelson and Jessica Brown, but I consider them both collaborators on this new edition. Early on, they embraced *The Curious Researcher* as a text for their writing classes, and they began sending me encouraging letters along with even more encouraging student essays. Two of those essays appear in the appendices of the book. Rachel and Jessica teach my own book better than I do, and for the many ways they've instructed me about how to teach research writing, I'm deeply grateful.

I have known Barry Lane for many years. He is a wonderful friend, writer, and teacher. In many ways, Barry is a co-author of everything I do. His praise of this book means a great deal to me, and our conversations about it continue to make it better. I look forward to more books, and more conversations about them, with this good friend.

Other writers and writing teachers who have contributed directly to *The Curious Researcher* or given me valuable ideas include Robin Lent, Brock Dethier, Barbara Tindall, Donna Qualley, Nora and Tony Nevin, Diane McAnaney, Sue Hertz, Jane Harrigan, Tamara Niedlsowski, Greg Bowe, Bronwyn Williams, Lad Tobin, Karen Uehling, Barbara Valdez, Patricia Sullivan, and Leaf Seligman.

I am also grateful to my editor at Allyn and Bacon, Joseph Opiela, who believed in this project from the very beginning, and also to his talented assistant, Kate Tolini. Susan Freese did a masterful job reining in my sometimes unruly prose.

I would like to thank those individuals who reviewed earlier versions of this book for Allyn and Bacon. Reviewers for the first edition included the following: Joseph T. Barwick, Central Piedmont Community College; Arnold J. Bradford, Northern Virginia Community College; Jack Branscomb, East Tennessee State University; Patricia E. Connors, Memphis State University; A. Cheryl Curtis, University of Hartford; John Fugate, J. Sargeant Reynolds Community College; Walter S. Minot, Gannon University; Michele Moragne e Silva, St. Edwards University; Al Starr, Essex Community College; Henrietta S. Twining, Alabama A&M; and Matthew Wilson, Rutgers University. And reviewers for the second edition included these individuals: Anne Maxham-Kastrinos, Washington State University; Martha W. Sipe, Boise State University; Sally Terrell, University of Hartford; and David Wasser, University of Hartford.

And finally, I am most indebted to my wife, Karen Kelley, who helped me see this project through during a difficult time in our lives.

INTRODUCTION

□ ■ □

Rethinking the Research Paper

Unlike most textbooks, this one begins with your writing, not mine. Find a fresh page in your notebook, grab a pen, and spend ten minutes doing the following exercise.

❑ *EXERCISE 1*
Collecting Golf Balls on Driving Ranges and Other Reflections

Most of us were taught to think before we write, to have it all figured out in our heads before we pick up our pens. This exercise asks you to think *through* writing rather than *before,* letting the words on the page lead you to what you want to say. With practice, that's surprisingly easy using a technique called *fastwriting.* Basically, you just write down whatever comes into your head, not worrying about whether you're being eloquent, grammatical, or even very smart. It's remarkably like talking to a good friend, not trying to be brilliant and even blithering a bit, but along the way discovering what you think. If the writing stalls, write about that, or write about what you've already written until you find a new trail to follow. Just keep your pen moving.

Step 1: Listed on the next page is a series of statements abo research papers and research. Choose one statement as a sta place for a five-minute fastwrite. First, through writing, thin whether you agree or disagree with the statement, and then explore why. Whenever possible, write about your own experience (anecdotes, stories, scenes, moments, etc.) with research and research papers as a means of thinking about what you believe about them.

> Researchers must always be objective.
>
> Research is an act of discovery.
>
> Writing in the first person—as *I*—in a research paper is always a bad idea.
>
> When I write a research paper, I have to say something original.
>
> Research mostly involves going to the library, collecting information from books and magazines, and putting it in your paper.
>
> I'm supposed to express my own opinions in a research paper.
>
> The process of research is basically finding facts that agree with my opinion.

Step 2: Now, consider the truth of some other statements, listed below. These statements have less to do with research papers than with how you see facts, information, and knowledge and how they're created. As in Step 1, choose one of these statements* to launch a five-minute fastwrite. Don't worry if you end up thinking about more than one statement in your writing. Like before, start by writing about whether you agree or disagree with the statement, and then explore why. Continually look for concrete connections between what you think about these statements and what you've seen or experienced in your own life.

> There is a big difference between facts and opinions.
>
> Pretty much everything you read in textbooks is true.
>
> People are entitled to their own opinions, and no one opinion is better than another.
>
> There's a big difference between *a fact* in the sciences and *a fact* in the humanities.
>
> How much I get out of school depends on the quality of my teachers.
>
> When two experts disagree, one of them has to be wrong.
>
> A story that doesn't have an ending isn't a very good story.

*Partial source for this list is Marlene Schommer, "Effects of Beliefs about the Nature of Knowledge," *Journal of Educational Psychology* 82 (1990): 498–504.

Very few of us recall the research papers we wrote in high school, and if we do, what we remember is not what we learned about our topics but what a bad experience it was. Joe was an exception. "I remember one assignment was to write a research paper on a problem in the world, such as acid rain, and then come up with your own solutions and discuss moral and ethical aspects of your solution, as well. It involved not just research but creativity and problem solving and other stuff."

For the life of me, I can't recall a single research paper I wrote in high school, but like Joe, I remember the one that I finally enjoyed doing a few years later in college. It was a paper on the whaling industry, and what I remember best was the introduction. I spent a lot of time on it, describing in great detail exactly what it was like to stand at the bow of a Japanese whaler, straddling an explosive harpoon gun, taking aim, and blowing a bloody hole in a humpback whale.

I obviously felt pretty strongly about the topic.

Unfortunately, many students feel most strongly about getting their research papers over with. So it's not surprising that when I tell my Freshman English course that one of their writing assignments will be an eight- to ten-page research paper, there is a collective sigh. They knew it was coming. For years, their high school teachers prepared them for the College Research Paper, and it loomed ahead of them as one of the torturous things you must do, a five-week sentence of hard labor in the library, picking away at cold, stony facts. Not surprisingly, students' eyes roll in disbelief when I add that many of them will end up liking their research papers better than anything they've written before.

I can understand why Joe was among the few in the class inclined to believe me. For many students, the library is an alien place, a wilderness to get lost in, a place to go only when forced. Others carry memories of research paper assignments that mostly involved taking copious notes on index cards, only to transfer pieces of information into the paper, sewn together like patches of a quilt. There seemed little purpose to it. "You weren't expected to learn anything about yourself with the high school research paper," wrote Jenn, now a college freshman. "The best ones seemed to be those with the most information. I always tried to find the most sources, as if somehow that would automatically make my paper better than the rest." For Jenn and others like her, research was a mechanical process and the researcher a lot like those machines that collect golf balls at driving ranges. You venture out to pick up information here and there, and then deposit it between the title page and the bibliography for your teacher to take a whack at.

GOOD RESEARCH AND GOOD WRITING

It should be no surprise that student research papers can be lifeless things. Many students believe that the quickest way to kill a piece of writing is to use *facts* and that there's nothing at all *creative* about the research process. Many instructors must believe this, too, since they often dread assigning research papers and then dread reading them. Fortunately, that's changing. More and more college instructors—and their students—are beginning to recognize this simple fact: *Good research doesn't have to mean bad writing.*

The deadline for your paper is around five weeks away. Though you have some hard work ahead, I do hope that by the time you finish your paper and this book, you will see the research process at least a little differently. I hope you will see the library as an ally, not an adversary, and not as a wilderness but familiar territory where you know how to find what you need. I hope you will see the research paper as a piece of writing that has as much to do with what you think and how you see things as the personal essay you wrote in English or the poem you jotted down in your journal last night. I hope you will see that a research paper can be engaging to read as well as to write and that facts are another kind of detail that, when used purposefully, can bring research writing to life. Maybe most important, I hope you will see the research paper as a means of satisfying your own curiosity, a curiosity that will drive you to dig deeper into your topic, even late at night when the library is deserted and you have a 7 A.M. class the next day.

USING THIS BOOK

The Exercises

Throughout this book, you'll be asked to do exercises that either help you prepare your research paper or actually help you write it. You'll need a research notebook in which you'll do the exercises and perhaps compile your notes for the paper. Any notebook will do, as long as there are sufficient pages and left margins. Your instructor may ask you to hand in the work you do in response to the exercises, so it would be useful to use a notebook with detachable pages.

Several of the exercises in this book ask that you use techniques like fastwriting and brainstorming. This chapter began with one, so you've already had a little practice with the two methods. Both fastwriting and brainstorming ask that you suspend judgment until you see what you come up with. That's pretty hard for most of us because we are so quick to criticize ourselves, particularly about writing. But if you can learn to get comfortable with the sloppiness that comes with writing almost as fast as you think, not bothering about grammar or punctuation, then you will be rewarded with a new way to think, letting your own words lead you in sometimes surprising directions. Though these so-called creative techniques seem to have little to do with the serious business of research writing, they can actually be an enormous help throughout the process. Try to ignore that voice in your head that wants to convince you that you're wasting your time using fastwriting or brainstorming. When you do, they'll start to work for you.

The Five-Week Plan

But more about creative techniques later. You have a research paper assignment to do. If you're excited about writing a research paper, that's great. You probably already know that it can be interesting work. But if you're dreading the work ahead of you, then your instinct might be to procrastinate, put it off until the week it's due. That would be a mistake, of course. If you try to rush through the research and the writing, you're absolutely guaranteed to hate the experience and add this assignment to the many research papers in the garbage dump of your memory. It's also much more likely that the paper won't be very good. Because procrastination is the enemy, this book was designed to help you budget your time and move through the research and writing process in five weeks. It may take you a little longer, or you may be able to finish your paper a little more quickly. But at least initially, use the book sequentially.

This book can also be used as a reference to solve problems as they arise. For example, suppose you're having a hard time finding enough information on your topic or you want to know how to plan for an interview. Use the Table of Contents as a key to typical problems and where in the book you can find some practical help with them.

Alternatives to the Five-Week Plan

Though *The Curious Researcher* is structured by weeks, you can easily ignore that plan and use the book to solve problems as they arise. This new edition of the book makes doing so even easier. The

"Contents by Subject" in the front of the text is keyed to a range of typical problems that arise for researchers: how to find a topic, how to focus a paper, how to handle a thesis, how to search the Internet, how to organize the material, how to take useful notes, and so on. Also new to this edition are the overviews of Modern Language Association (MLA) and American Psychological Association (APA) research paper conventions in Appendixes A and B, respectively. These appendixes provide complete guides to both formats and make it easier to find answers to your specific technical questions at any point in the process of writing your paper.

THE RESEARCH PAPER AND THE RESEARCH REPORT

Anyone who spent a few years in high school clutching index cards and making a beeline for the *Encyclopaedia Britannica* or *Reader's Guide* every time a research paper assignment was given will probably be glad that the college-level research paper will be a different experience. It's a little hard to get excited about paraphrasing an encyclopedia or the December issue of *Time* onto notecards and then inserting that information into the paper. But that's what seemed the logical thing to do when the assignment was to write a paper that reflects what's known about your topic. That's called a *research report,* and it's a fairly common assignment in the first few years of high school.

Discovering Your Purpose

For the paper you're about to write, the information you collect must be used much more *purposefully* than simply reporting what's known about a particular topic. Most likely, you will define what that purpose is. For example, you may end up writing a paper whose purpose is to argue a point—say, eating meat is morally suspect because of the way stock animals are treated at slaughterhouses. Or your paper's purpose may be to reveal some less known or surprising aspect of a topic—say, how the common housefly's eating habits are not unlike our own. Or your paper may set out to explore a thesis, or idea, that you have about your topic—for example, your topic is the cultural differences between men and women, and you suspect the way girls and boys play as children reflects the social differences evident between the genders as adults.

Whatever the purpose of your paper turns out to be, the process usually begins with something you've wondered about, some itchy question about an aspect of the world you'd love to know the answer to. It's the writer's curiosity—not the teacher's—that is at the heart of the college research paper.

In some ways, frankly, *research reports* are easier. You just go out and collect as much stuff as you can, write it down, organize it, and write it down again in the paper. Your job is largely mechanical and often deadening. In the *research paper,* you take a much more active role in *shaping and being shaped by* the information you encounter. That's harder because you must evaluate, judge, interpret, and analyze. But it's also much more satisfying because what you end up with says something about who you are and how you see things.

HOW FORMAL SHOULD IT BE?

When I got a research paper assignment, it often felt as if I was being asked to change out of blue jeans and a wrinkled oxford shirt and get into a stiff tuxedo. Tuxedos have their place, such as at the junior prom or the Grammy Awards, but they're just not me. When I first started writing research papers, I used to think that I *had* to be formal, that I needed to use big words like *myriad* and *ameliorate* and to use the pronoun *one* instead of *I.* I thought the paper absolutely needed to have an introduction, body, and conclusion—say what I was going to say, say it, and say what I said. It's no wonder that the first college research paper I had to write—on Plato's *Republic* for a philosophy class—seemed to me as though it were written by someone else. I felt at arm's length from the topic I was writing about.

You may be relieved to know that not all research papers are necessarily rigidly formal or dispassionate. Some are. Research papers in the sciences, for example, often have very formal structures, and the writer seems more a reporter of results than someone who is passionately engaged in making sense of them. This *formal stance* puts the emphasis where it belongs: on the validity of the data in proving or disproving something, rather than on the writer's individual way of seeing something. Some papers in the social sciences, particularly scholarly papers, take a similarly formal stance, where the writer not only seems invisible but seems to have little relation to the subject. There are many reasons for this approach. One is that *objectivity*—or as one philosopher put it, "the separation of the perceiver from the thing perceived"—is traditionally a highly valued principle among some scholars and researchers. For example, if I'm writing a paper on the effectiveness of Alcoholics Anonymous (AA), and I confess that my father—who attended AA—drank himself to death, can I be trusted to see things clearly?

Yes, *if* my investigation of the topic seems thorough, balanced, and informative. And I think it may be an even better paper because my passion for the topic will encourage me to look at it more closely. However, many scholars these days are openly skeptical about claims of objectivity. Is it really possible to separate the perceiver from the thing perceived? If nothing else, aren't our accounts of reality always mediated by the words we use to describe it? Can language ever be objective? Though the apparent impersonality of their papers may suggest otherwise, most scholars are not nearly as dispassionate about their topics as they seem. They are driven by the same thing that will send you to the library over the next few weeks—their own curiosity—and most recognize that good research often involves both objectivity and subjectivity. As the son of an alcoholic, I am motivated to explore my own perceptions of his experience in AA, yet I recognize the need to verify those against the perceptions of others with perhaps more knowledge.

Your instructor may want you to write a formal research paper. You should determine if a formal paper is required when you get the assignment. (See the box "Questions to Ask Your Instructor about the Research Assignment.") Also make sure that you understand what the word *formal* means. Your instructor may have a specific format you should follow or tone you should keep. But more likely, she is much more interested in your writing a paper that reflects some original thinking on your part and that is also lively and interesting to read. Though this book will help you write a formal research paper, it encourages what might be called a *research essay,* a paper that does not have a prescribed form, though it is as carefully researched and documented as a more formal paper.

"ESSAYING" OR ARGUING?

Essay is a term that is used so widely to describe school writing that it often doesn't seem to carry much particular meaning. But I have something particular in mind.

The term *essai* was coined by Michel Montaigne, a sixteenth-century Frenchman; in French, it means "to attempt" or "to try." For Montaigne and the essayists who follow his tradition, the essay is less an opportunity *to prove* something than an attempt *to find out.* An essay is often exploratory rather than argumentative, testing the

QUESTIONS TO ASK YOUR INSTRUCTOR ABOUT THE RESEARCH ASSIGNMENT

It's easy to make assumptions about what your instructor expects for the research paper assignment. After all, you've probably written such a paper before and may have had the sense that the "rules" for doing so were handed down by God. Unfortunately, those assumptions may get in the way of writing a good paper, and sometimes they're dead wrong. If you got a handout describing the assignment, it may answer the questions below, but if not, make sure you raise them with your instructor when he gives the assignment.

- How would you describe the audience for this paper?
- Do you expect the paper to be in a particular form or organized in a special way? Or can I develop a form that suits the purpose of my paper?
- Do you have guidelines about format (margins, title page, outline, bibliography, citation method, etc.)?
- Can I use other visual devices (illustrations, subheadings, bulleted lists, etc.) to make my paper more readable?
- Can I use the pronoun *I* when appropriate?
- Can my own observations or experiences be included in the paper if relevant?
- Can I include people I interview as sources in my paper? Would you encourage me to use "live" sources as well as published ones?
- Should the paper *sound* a certain way, have a particular tone, or am I free to use a writing voice that suits my subject and purpose?

truth of an idea or attempting to discover what might be true. (Montaigne even once had coins minted that said *Que scais-je*—"What do I know?") The essay is often openly subjective and frequently takes a conversational, even intimate, form.

Now, this probably sounds nothing like any research paper you've ever written. Certainly, the dominant mode of the academic research paper is impersonal and argumentative. But if you consider writing a *research essay* instead of the usual *research paper,* four things might happen:

1. *You'll discover your choice of possible topics suddenly expands.* If you're not limited to arguing a position on a topic, then you can explore any topic that you find puzzling in interesting ways and you can risk asking questions that might complicate your point of view.

2. *You'll find that you'll approach your topics differently.* You'll be more open to conflicting points of view and perhaps more willing to change your mind about what you think. As one of my students once told me, this is a more honest kind of objectivity.

3. *You'll see a stronger connection between this assignment and the writing you've done all semester.* Research is something all writers do, not a separate activity or genre that exists only upon demand. You may discover that research can be a revision strategy for improving essays you wrote earlier in the semester.

4. *You'll find that you can't hide.* The research report often encourages the writer to play a passive role; the research essay doesn't easily tolerate Joe Fridays ("Just the facts, ma'am") easily. You'll probably find this both liberating and frustrating. While you may likely welcome the chance to incorporate your opinions, you may find it difficult to add your voice to those of your sources.

You may very well choose to write a paper that argues a point for this assignment (and, by the way, even an essay has a point). After all, the argumentative paper is the most familiar form of the academic research paper. In fact, a sample research paper that uses argument is featured in Appendix B. It's an interesting, well-researched piece in which the writer registers a strong and lively presence. But I hope you might also consider essaying your topic, an approach that encourages a kind of inquiry that may transform your attitudes about what it means to write research.

BECOMING AN AUTHORITY BY USING AUTHORITIES

Whether formal or less so, all research papers attempt to be *authoritative.* That is, they rely heavily on a variety of credible sources beyond the writer who helped shape the writer's point of view. Those sources are mostly already published material, but they can also be other people, usually experts in relevant fields whom you interview for their perspectives. Don't underestimate the value of "live" and other nonlibrary sources. Authorities don't just live in books. One might live in the office next door to your class or be easily accessible through the Internet.

Though in research papers the emphasis is on using credible outside sources, that doesn't mean that your own experiences or observations should necessarily be excluded from your paper when they're relevant. In fact, in some papers, they are essential. For example, if you decide to write a paper on Alice Walker's novel *The Color Purple,* your own reading of the book—what strikes you as important—should be at the heart of your essay. Information from literary critics you discover in your research will help you develop and support the assertions you're making about the novel. That support from people who are considered experts—that is, scholars, researchers, critics, and practitioners in the field you're researching—will rub off on you, making your assertions more convincing, or authoritative.

Reading and talking to these people will also change your thinking, which is part of the fun of research. You will actually learn something, rather than remain locked into preconceived notions.

"It's Just My Opinion"

In the end, *you* will become an authority of sorts. I know that's hard to believe. One of the things my students often complain about is their struggle to put their opinions in their papers: "I've got all these facts, and sometimes I don't know what to say other than whether I disagree or agree with them." What these students often *seem* to say is that they don't really trust their own authority enough to do much more than state briefly what they feel: "Facts are facts. How can you argue with them?"

Step 2 of the Exercise that began this chapter (see page 2) may have started you thinking about these questions. I hope the research assignment you are about to start keeps you thinking about your beliefs about the nature of knowledge. Are facts unassailable? Or are they simply claims that can be evaluated like any others? Is the struggle to evaluate conflicting claims an obstacle to doing research or the point of it? Are experts supposed to know all the answers? What makes one opinion more valid than another? What makes *your* opinion valid?

I hope you write a great essay in the next five or so weeks. But I also hope that the process you follow in doing so inspires you to reflect on how you—and perhaps all of us—come to know what seems to be true. I hope you find yourself doing something you may not have done much before: thinking about thinking.

FACTS DON'T KILL

You probably think the words *research paper* and *interesting* are mutually exclusive. The prevalent belief among my students is that the minute you start having to use facts in your writing, then the prose wilts and dies like an unwatered begonia. It's an understandable attitude. There are many examples of informational writing that is dry and wooden, and among them, unfortunately, may be some textbooks you are asked to read for other classes.

But factual writing doesn't have to be dull. You may not consider the article "Why God Created Flies" (see the following exercise) a research paper. It may be unlike the research papers you've been assigned in some ways; it has no citation of sources, a heavy use of the author's personal experiences, and an informal structure. But it is a piece that's research based, containing plenty of facts about flies. It has a clear purpose and a thesis. Yet despite that, it's a pretty good read.

❑ *EXERCISE 2*
Bringing "Flies" to Life

Read Conniff's "Why God Created Flies" first for pure enjoyment. Then reread the article with your pen in hand, and, in your research notebook, list what you think makes the article interesting. Be specific. What devices or techniques does Conniff use to hold your interest? Your instructor may want you to bring your list to class for discussion.

Other questions for class discussion:

- In what ways is this article unlike any research paper you've ever written?
- How does Conniff handle the factual material? How is it woven into the article?
- What kinds of facts does he use? Are they memorable? Why?
- What is Conniff's main point, or thesis? Where is it stated most clearly?

- Does the humor in the article make it less authoritative? Does the author's use of personal experiences and observations strengthen or weaken the research?
- Is this article *objective?* In what ways? In what ways is it *subjective?* Can a good research paper be both without sacrificing its authority?

□ ■ □

Why God Created Flies

by Richard Conniff

THOUGH I HAVE been killing them for years now, I have never tested the folklore that, with a little cream and sugar, flies taste very much like black raspberries. So it's possible I'm speaking too hastily when I say there is nothing to like about houseflies. Unlike the poet who welcomed a "busy, curious, thirsty fly" to his drinking cup, I don't cherish them for reminding me that life is short. Nor do I much admire them for their function in clearing away carrion and waste. It is, after all, possible to believe in the grand scheme of recycling without liking undertakers.

A fly is standing on the rim of my beer glass as I write these words. Its vast, mosaic eyes look simultaneously lifeless and mocking. It grooms itself methodically, its forelegs entwining like the arms of a Sybarite luxuriating in bath oil. Its hind legs twitch across the upper surface of its wings. It pauses, well fed and at rest, to contemplate the sweetness of life.

We are lucky enough to live in an era when scientists quantify such things, and so as I type and wait my turn to drink, I know that the fly is neither busy nor curious. The female spends 40.6 percent of her time doing nothing but contemplating the sweetness of life. I know that she not only eats unspeakable things, but spends an additional 29.7 percent of her time spitting them back up and blowing bubbles with her vomit. The male is slightly less assiduous at this deplorable pastime but also defecates on average every four and a half minutes. Houseflies seldom trouble us as a health threat anymore, at least in this coun-

Reprinted by permission from Conniff, Richard, *Spineless Wonders—Strange Tales of the Invertebrae World* (Henry Holt & Co., 1996).

try, but they are capable of killing. And when we are dead (or sooner, in some cases), they dine on our corrupted flesh.

It is mainly this relentless intimacy with mankind that makes the housefly so contemptible. Leeches or dung beetles may appall us, but by and large they satisfy their depraved appetites out of our sight. Houseflies, on the other hand, routinely flit from diaper pail to dinner table, from carrion to picnic basket. They are constantly among us, tramping across our food with God-knows-what trapped in the sticky hairs of their half-dozen legs.

Twice in this century, Americans have waged war against houseflies, once in a futile nationwide "swat the fly" campaign and again, disastrously, with DDT foggings after World War II. The intensity of these efforts, bordering at times on the fanatic, may bewilder modern Americans. "Flies or Babies? Choose!" cried a headline in the *Ladies' Home Journal* in 1920. But our bewilderment is not due entirely to greater tolerance or environmental enlightenment. If we have the leisure to examine the fly more rationally now, it is primarily because we don't suffer its onslaughts as our predecessors did. Urban living has separated us from livestock, and indoor plumbing has helped us control our own wastes and thus control houseflies. If that changed tomorrow, we would come face to face with the enlightened, modern truth: With the possible exception of *Homo sapiens,* it is hard to imagine an animal as disgusting or improbable as the housefly. No bestiary concocted from the nightmares of the medieval mind could have come up with such a fantastic animal. If we want to study nature in its most exotic permutations, the best place to begin is here, at home, on the rim of my beer glass.

IN THIS COUNTRY, more than a dozen fly species visit or live in the house. It is possible to distinguish among some of them only by such microscopic criteria as the pattern of veins in the wings, so all of them end up being cursed as houseflies. Among the more prominent are the blue- and greenbottle flies, with their iridescent abdomens, and the biting stable flies, which have served this country as patriots, or at least provocateurs. On July 4, 1776, their biting encouraged decisiveness among delegates considering the Declaration of Independence. "Treason," Thomas Jefferson wrote, "was preferable to discomfort."

The true housefly, *Musca domestica,* does not bite. (You may think this is something to like about it, until you find out what it does instead.) *M. domestica,* a drab fellow of salt-and-pepper complexion, is the world's most widely distributed insect species and probably the most familiar, a status achieved through its pronounced fondness for breeding in pig, horse, or human excrement. In choosing at some point in the immemorial past to concentrate on the wastes around human habitations, *M. domestica* made a major career move. Bernard Greenberg of the University of Illinois at Chicago has traced human representations of the housefly back to a Mesopotamian cylinder seal from 3000 B.C. But houseflies were probably with us even before we had houses, and they spread with human culture.

Like us, the housefly is prolific, opportunistic, and inclined toward exploration. It can adapt to a diet of either vegetables or meat, preferably somewhat ripe. It will lay its eggs not just in excrement but in rotting lime peels, birds nests, carrion, even flesh wounds that have become infected and malodorous. Other flies aren't so flexible. For instance, *M. autumnalis,* a close relative, prefers cattle dung and winds up sleeping in pastures more than in houses or yards.

Although the adaptability and evolutionary generalization of the housefly may be admirable, they raise one of the first great questions about flies: Why is there this dismaying appetite for abomination?

Houseflies not only defecate constantly but do so in liquid form, which means they are in constant danger of dehydration. The male can slake his thirst and get most of the energy he needs from nectar. But fresh manure is a good source of water, and it contains the dissolved protein the female needs to make eggs. She also lays her eggs in excrement or amid decay so that when the maggots hatch, they'll have a smorgasbord of nutritious microorganisms on which to graze.

Houseflies bashing around the kitchen or the garbage shed thus have their sensors attuned to things that smell sweet, like flowers or bananas, and to foul-smelling stuff like ammonia and hydrogen sulfide, the products of fermentation and putrefaction. (Ecstasy for the fly is the stinkhorn fungus, a source of sugar that smells like rotting meat.)

The fly's jerky, erratic flight amounts to a way of covering large territories in search of these scents, not just for food but for romance and breeding sites. Like dung beetles and other flying insects, the fly will zigzag upwind when it gets a whiff of something good (or, more often, bad) and follow the scent plume to its source.

HENCE THE SECOND question about the housefly: How does it manage to fly so well? And the corollaries: Why is it so adept at evading us when we swat it? How come it always seems to land on its feet, usually upside-down on the ceiling, having induced us to plant a fist on the spot where it used to be, in the middle of the strawberry trifle, which is now spattered across the tablecloth, walls, loved ones, and honored guests?

The housefly's manner of flight is a source of vexation more than wonder. When we launch an ambush as the oblivious fly preens and pukes, its pressure sensors alert it to the speed and direction of the descending hand. Its wraparound eyes are also acutely sensitive to peripheral movement, and they register changes in light about ten times faster than we do. (A movie fools the gullible human eye into seeing continuous motion by showing it a sequence of twenty-four still pictures a second. To fool a fly would take more than 200 frames a second.) The alarm flashes directly from the brain to the middle set of legs via the largest, and therefore fastest, nerve fiber in the body. This causes so-called starter muscles to contract, simultaneously revving up the wing muscles and pressing down the middle legs, which catapult the fly into the air.

The fly's wings beat 165 to 200 times a second. Although this isn't all that fast for an insect, it's more than double the wingbeat of the speediest hummingbird and about twenty times faster than any repetitious movement the human nervous system can manage. The trick brought off by houseflies and many other insects is to remove the wingbeat from direct nervous system control, once it's switched on. Two systems of muscles, for upstroke and downstroke, are attached to the hull of the fly's midsection, and they trigger each other to work in alternation. When one set contracts, it deforms the hull, stretching the other set of muscles and making it contract automatically a fraction of a second later. To keep this seesaw rhythm going, openings in the midsection stoke the muscles with oxygen that comes directly from the outside (flies have no lungs). Meanwhile the fly's blood (which lacks hemoglobin and is therefore colorless) carries fuel to the cells fourteen times faster than when a fly is at rest. Flies can turn a sugar meal into useable energy so fast that an exhausted fly will resume flight almost instantly after eating. In humans . . . but you don't want to know how ploddingly inadequate humans are by comparison.

An airborne fly's antennae, pointed down between its eyes, help regulate flight, vibrating in response to airflow. The fly also uses a set of stubby wings in back, called halteres, as a gyroscopic device. Flies are skillful at veering and dodging—it sometimes seems that they are doing barrel rolls and Immelmann turns to amuse themselves while we flail and curse. But one thing they cannot do is fly upside-down to land on a ceiling. This phenomenon puzzled generations of upward-glaring, strawberry-trifle-drenched human beings, until high-speed photography supplied the explanation. The fly approaches the ceiling right-side up, at a steep angle. Just before impact, it reaches up with its front limbs, in the manner of Superman exiting a telephone booth for takeoff. As these forelegs get a grip with claws and with the sticky, glandular hairs of the footpads, the fly swings its other legs up into position. Then it shuts down its flight motor, out of swatting range and at ease.

While landing on the ceiling must be great fun, humans tend to be more interested in what flies do when they land on food. To find out, I trapped the fly on the rim of my beer glass. (Actually, I waited till it found a less coveted perch, then slowly lowered a mayonnaise jar over it.) I'd been reading a book called *To Know a Fly,* in which author Vincent Dethier describes a simple way of seeing how the fly's proboscis works. First I refrigerated the fly to slow it down and anesthetize it. Then I attempted to attach a thin stick to its wing surface with the help of hot candlewax. It got away. I brought it back and tried again. My four-year-old son winced and turned aside when I applied the wax. "I'm glad I'm not a fly," he said, "or you might do that to me." I regarded him balefully but refrained from mentioning the ant colony he had annihilated on our front walk.

Having finally secured the fly, I lowered its feet into a saucer of water. Flies have taste buds in their feet, and when they walk on something good (bad), the proboscis, which is normally folded up neatly inside the head, automatically flicks down. No response. I added sugar to the water, an irresistible combination. Nothing. More sugar. Still nothing. My son wandered off, bored. I apologized to the fly, killed it, and decided to look up the man who had put me in the awkward position of sympathizing with a fly, incidentally classing me in my son's eyes as a potential war criminal.

DETHIER, A BIOLOGIST at the University of Massachusetts at Amherst, turned out to be a gentle, deferential fellow in his mid-seventies, with weathered, finely wrinkled skin and gold-rimmed oval eyeglasses on a beak nose. He suggested mildly that the fly might not have responded because it was outraged at the treatment it received. It may also have eaten recently, or it may have been groggy from hibernation. (Some flies sit out the winter in diapause, in which hormones induce inactivity in response to shortened day length. But cold, not day length, is what slows down hibernating species like the housefly, and the sudden return of warmth can start them up again. This is why a fly may miraculously take wing on a warm December afternoon in the space between my closed office window and the closed storm window outside, a phenomenon I had formerly regarded as new evidence for spontaneous generation.)

Dethier has spent a lifetime studying the fly's sense of taste, "finding out where their tongues and noses are, as it were." He explained the workings of the proboscis to me.

Fly taste buds, it seems, are vastly more sensitive than ours. Dethier figured this out by taking saucers of water containing steadily decreasing concentrations of sugar. He found the smallest concentration a human tongue could taste. Then he found the smallest concentration that caused a hungry fly to flick out its proboscis. The fly, with 1,500 taste hairs arrayed on its feet and in and around its mouth, was ten million times more sensitive.

When the fly hits pay dirt, its proboscis telescopes downward and the fleshy lobes at the tip puff out. These lips can press down tight to feed on a thin film of liquid, or they can cup themselves around a droplet. They are grooved crosswise with a series of parallel gutters, and when the fly starts pumping, the liquid is drawn up through these gutters. Narrow zigzag openings in the gutters filter the food, so that even when it dines on excrement, the fly can "choose" some morsels and reject others.

A drop of vomit may help dissolve the food, making it easier to lap up. Scientists have also suggested that the fly's prodigious vomiting may be a way of mixing enzymes with the food to aid digestion.

If necessary, the fly can peel its lips back out of the way and apply its mouth directly to the object of its desire. Although a housefly does not have true teeth, its mouth is lined with a jagged, bladelike edge that is useful for scraping. In his book *Flies and Disease,* Bernard Greenberg writes that some blowflies (like the one on the rim of my beer glass, which turned out to be a *Phormia regina*) "can bring 150 teeth into action, a rather effective scarifier for the superficial inoculation of the skin, conjunctiva, or mucous membranes."

HENCE THE FINAL great question about flies: What awful things are they inoculating us with when they flit across our food or land on our sleeping lips to drink our saliva? Over the years, authorities have suspected flies of spreading more than sixty diseases, from diarrhea to plague and leprosy. As recently as 1951, the leading expert on flies repeated without demurring the idea that the fly was "the most dangerous insect" known, a remarkable assertion in a world that also includes mosquitoes. One entomologist tried to have the housefly renamed the "typhoid fly."

The hysteria against flies earlier in the century arose, with considerable help from scientists and the press, out of the combined ideas that germs cause disease and that flies carry germs. In the Spanish-American War, easily ten times as many soldiers died of disease, mostly typhoid fever, as died in battle. Flies were widely blamed, especially after a doctor observed particles of lime picked up in the latrines still clinging to the legs of flies crawling over army food. A British politician argued that flies were not "dipterous angels" but "winged sponges speeding hither and thither to carry out the foul behests of Contagion." American schools started organizing "junior sanitary police" to point the finger at fly breeding sites. Cities sponsored highly publicized "swat the fly" campaigns. In Washington, D.C., in 1912, a consortium of children killed 343,800 flies and won a $25 first prize. (This is a mess of flies, 137.5 swatted for every penny in prize money, testimony to the slowness of summers then and the remarkable agility of children—or perhaps to the overzealous imagination of contest sponsors. The figure does not include the millions of dead flies submitted by losing entrants.)

But it took the pesticide DDT, developed in World War II and touted afterwards as "the killer of killers," to raise the glorious prospect of "a flyless millennium." The fly had by then been enshrined in the common lore as a diabolical killer. In one of the "archy and mehitabel" poems by Don Marquis, a fly visits garbage cans and sewers to "gather up the germs of typhoid, influenza, and pneumonia on my feet and wings" and spread them to humanity, declaring that "it is my mission to help rid the world of these wicked persons/i am a vessel of righteousness."

Public health officials were deadly serious about conquering this arch fiend, and for them DDT was "a veritable godsend." They recommended that parents use wallpaper impregnated with DDT in nurseries and playrooms to protect children. Believing that flies spread infantile paralysis, cities suffering polio epidemics frequently used airplanes to fog vast areas with DDT. Use of the chemical actually provided some damning evidence against flies, though not in connection with polio. Hidalgo County in Texas, on the Mexican border, divided its towns into two groups and sprayed one with DDT to eliminate flies. The number of children suffering and dying from acute diarrheal infections caused by *Shigella* bacteria declined in the sprayed areas but remained the same in the unsprayed zones. When DDT spraying was stopped in the first group and switched to the second, the dysentery rates began to reverse. Then the flies developed resistance to DDT, a small hitch in the godsend. In state parks and vacation spots, where DDT had provided relief from the fly nuisance, people began to notice that songbirds were also disappearing.

IN THE END, the damning evidence was that we were contaminating our water, ourselves, and our affiliated population of flies with our own filth (not to mention DDT). Given access to human waste through inadequate plumbing or sewerage treatment, flies can indeed pick up an astonishing variety of pathogens. They can also reproduce at a godawful rate: In one study, 4,042 flies hatched from a scant shovelful, one-sixth of a cubic foot, of buried night soil. But whether all those winged sponges can transmit the contaminants they pick up turns out to be a tricky question, the Hidalgo County study being one of the few clearcut exceptions. Of polio, for instance, Bernard Greenberg writes, "There is ample evidence that human populations readily infect flies. . . . But we are woefully ignorant whether and to what extent flies return the favor."

Flies probably are not, as one writer declared in the throes of hysteria, "monstrous" beings "armed with horrid mandibles . . . and dripping with poison." A fly's body is not, after all, a playground for microbes. Indeed, bacterial populations on its bristling, unlovely exterior tend to decline quickly under the triple threat of compulsive cleaning, desiccation, and ultraviolet radiation. (Maggots actually produce a substance in their gut that kills off whole populations of bacteria, which is one reason doctors have sometimes used them to clean out infected wounds.) The fly's "microbial cargo," to use Greenberg's phrase, reflects human uncleanliness. In one study, flies from a city neighborhood with poor facilities carried up to 500 million bacteria, while flies from a prim little suburb not far away yielded a maximum count of only 100,000.

But wait. While I am perfectly happy to suggest that humans are viler than we like to think, and flies less so, I do not mean to rehabilitate the fly. Any animal that kisses offal one minute and dinner the next is at the very least a social abomination. What I am coming around to is St. Augustine's idea that God created flies to punish human arrogance, and not just the calamitous technological arrogance of DDT. Flies are, as one biologist has remarked, the resurrection and the reincarnation of our own dirt, and this is surely one reason we smite them down with such ferocity. They mock our notions of personal grooming with visions of lime particles, night soil, and dog leavings. They toy with our delusions of immortality, buzzing in the ear as a memento mori (a researcher in Greenberg's lab assures me that flies can strip a human corpse back to bone in about a week, if the weather is fine). Flies are our fate, and one way or another they will have us.

It is a pretty crummy joke on God's part, of course, but there's no point in getting pouty about it and slipping into unhealthy thoughts about nature. What I intend to do, by way evening the score, is hang a strip of flypaper and cultivate the local frogs and snakes, which have a voracious appetite for flies (flycatchers don't, by the way; they seem to prefer wasps and bees). Perhaps I will get the cat interested, as a sporting proposition. Meanwhile, I plan to get a fresh beer and sit back with my feet up and a tightly rolled newspaper nearby. Such are the consolations of the ecological frame of mind.

I love "Why God Created Flies" partly because it never occurred to me when I first read it that I was reading research. Elmore Leonard, a distinguished fiction writer, says that when his writing *sounds* like writing, he needs to rewrite it. His prose—which is lean, efficient, yet powerful—reflects that philosophy. It doesn't call attention to itself. Much research writing does. It lumbers along from fact to fact and quote to quote, saying "Look at how much I know!" *Demonstrating* knowledge is not nearly as impressive as *using* it toward some end. Conniff does just that, masterfully weaving surprising information about the common housefly with his longing to determine where in God's plan the pest might fit in.

It's informative, it's funny, and yes, it's research. Richard Conniff is not a bug expert. (If he were, he'd probably be published in the *Journal of Entomology*.) He's just a guy who noticed a fly on his beer glass and wondered, What is it doing there? The best research often starts that way, with what at first seems to be a simple question. Like a lucky archaeologist, the researcher often finds more than he expects just below the surface of even the simplest questions. But first, he's got to want to dig.

1

□ ■ □

The First Week

THE IMPORTANCE OF GETTING CURIOUS

A few years back, I wrote a book about lobsters. At first, I didn't intend it to be a book. I didn't think there was that much to say about lobsters. But the more I researched the subject, the more questions I had and the more places I found to look for answers. Pretty soon, I had 300 pages of manuscript.

My curiosity about lobsters began one year when the local newspaper printed an article about what terrible shape the New England lobster fishery was in. The catch was down 30 percent, and the old-timers were saying it was the worst year they'd seen since the thirties. Even though I grew up in landlocked Chicago, I'd always loved eating lobsters after being introduced to them at age eight at my family's annual Christmas party. Many years later, when I read the article in my local newspaper about the vanishing lobsters, I was alarmed. I wondered, Will lobster go the way of caviar and become too expensive for people like me?

That was the question that triggered my research, and it soon led to more questions. What kept me going was my own curiosity. It's the same curiosity that motivated Richard Conniff to research and then write about the houseflies landing on the lip of his beer glass. If your research assignment is going to be successful, you need to get curious, too. If you're bored by your research topic, your paper will almost certainly be boring, as well, and you'll end up hating writing research papers as much as ever.

Learning to Wonder Again

Maybe you're naturally curious, a holdover from childhood when you were always asking Why? Or maybe your curiosity paled as you got older, and you forgot that being curious is the best reason for wanting to learn things. Whatever condition it's in, your curiosity must be the driving force behind your research paper. It's the most essential ingredient. The important thing, then, is this: *Choose your research topic carefully. If you lose interest in it, change your topic to one that does interest you or find a different angle.*

In most cases, instructors give students great latitude in choosing their research topics. (Some instructors narrow the field, asking students to find a focus within some broad, assigned subject. When the subject has been assigned, it may be harder for you to discover what you are curious about, but it won't be impossible, as you'll see.) Some of the best research topics grow out of your own experience (though they certainly don't have to), as mine did when writing about lobster overfishing or Conniff's did when writing about houseflies. Begin searching for a topic by asking yourself this question: What have I seen or experienced that raises questions that research can help answer?

Getting the Pot Boiling

A subject might bubble up immediately. For example, I had a student who was having a terrible time adjusting to her parents' divorce. Janabeth started out wanting to know about the impact of divorce on children and later focused her paper on how divorce affects father-daughter relationships.

Kim remembered spending a rainy week on Cape Cod with her father, wandering through old graveyards, looking for the family's ancestors. She noticed patterns on the stones and wondered what they meant. She found her ancestors as well as a great research topic.

Manuel was a divorced father of two, and both of his sons had recently been diagnosed with attention-deficit disorder (ADD). The boys' teachers strongly urged Manuel and his wife to arrange drug therapy for their sons, but they wondered whether there might be any alternatives. Manuel wrote a moving and informative research essay about his gradual acceptance of drug treatment as the best solution for his sons.

For years, Wendy loved J. D. Salinger's work but never had the chance to read some of his short stories. She jumped at the opportunity to spend five weeks reading and thinking about her favorite author. She later decided to focus her research paper on Salinger's notion of the misfit hero.

Sometimes other people, often unwittingly, will give you great topic ideas. Recently, a student in my composition class had a conversation with her friends in the wee hours of the morning about the Public Media Resource Center (PMRC), a Washington, D.C.–based group that is pushing for labeling music with pornographic or violent content. In the midst of the discussion, she realized how strongly she felt about the potential for censorship. She had a topic.

Sometimes, one topic triggers another. Chris, ambling by Thompson Hall, one of the oldest buildings on his campus, wondered about its history. After a little initial digging, he found some 1970 newsclips from the student newspaper, describing a student strike that paralyzed the school. The controversy fascinated him more than the building did, and he pursued the topic. He wrote a great paper.

If you're still drawing a blank, try the following exercise in your notebook.

❑ *EXERCISE 1.1*
What Do You Want to Know?

Make an "authority" list. In five minutes, brainstorm a list of things about which you already know something but would like to learn more. But first, a little more on how to make brainstorming work for you.

Essentially, you'll be making a fast list of whatever comes to mind. Don't censor yourself. Write down things even if they seem stupid. Don't worry if you build a list of words and phrases that only you can decipher; the list is meant exclusively for you. You'll probably find that your list comes in waves—you'll jot down four or five things, then draw a blank, and you'll think of four more. To help you find new waves to ride, write the following words across the top of a blank page in your notebook, and use them as column headings to inspire additional items for your list:

```
PLACES, THINGS, TRENDS, TECHNOLOGIES, PEOPLE,
CONTROVERSIES, HISTORY, JOBS, HABITS, HOBBIES,
DISCIPLINES
```

Relax and enjoy your wandering mind. Ready? Begin.

Now make a "nonauthority" list. Take another five minutes, and brainstorm a list of things you don't know much about but would like to learn more. Before you begin the first part of this exercise, refer again to the words you wrote across the top of your notebook page. Use them to trigger new items for this list, too.

Look at both lists, reviewing all the items you've jotted down. Circle one item you'd like to look at more closely, that makes you wonder.

Finally, take another five minutes and build a list of questions about your subject that you'd love to learn the answers to. Write down any questions that come to mind.

If you are getting nowhere with your subject, try another from the list. Repeat the last few steps with another subject until you produce a good list of questions. Save them for later.

Following are the lists I made recently when I did this exercise.

Things I know something about:

dinghy sailing
St. John
trout fishing
western Montana
Idaho desert
teaching writing
Joan Didion
Wallace Stegner
Robert Bly
environmental movement
woodstoves
finish carpentry
Earth Day
plant taxonomy
ecology
birdwatching
Great Lakes
alcoholism
Lake Michigan
lobsters
Maine coast
tying flies
keeping a journal
research paper
publishing industry
Alaskan pipeline
draft for Vietnam War
gardening
Robinson Jeffers
literary journalism
public relations
nonprofit organizations
Boundary Waters Canoe Area (BWCA)
direct mail fund-raising
sibling rivalry
psychotherapy
guitar

Things I don't know much about but would like to learn more:

fatherhood
masculinity
masculine myths
autobiographical memory
Joseph Campbell
Mary Hallock Foote
astronomy
building telescopes
New Zealand
Snake River Plain geology
cabinet making
restoration of Atlantic salmon
decline of salmon runs in northwest U.S.
trout fishing in New England
decline of mills in New England
labor history in the mills
International Workers of the World (IWW)
Oregeon Trail
offshore groundfish fishery
Piscataqua River
salvage logging
water use in the western U.S.
childhood development
Jean Piaget
macroeconomics

how the Federal Reserve works
operating large sailing ships
writing poetry
invertebrate zoology
roses
playwriting
art history
photography
rain forests
greenhouse effect
ice skating
Internet
Web pages
computers and writing

One item selected from both lists:

Great Lakes

Questions about my selected item:

Why do lake levels seem to fluctuate so mysteriously?

How did the Army Corp of Engineers reverse the direction of the Chicago River?

Do the lakes have a rich commercial fishing industry?

Where are most of the lakes' shipwrecks concentrated?

Where are the most dangerous waters?

Is Lake Superior still pristine?

What distinguishes the lakes from each other?

Which lakes are most polluted?

What's the nature of the pollution?

Is it true that fish caught in most of the lakes should not be eaten because of toxins in their fatty tissues?

How did the St. Lawrence Seaway and canals connecting the lakes to the sea affect the ecology?

Did the transplantation of salmon in the late 1960s affect the ecology?

What happened to the massive alewife die-offs I remember hearing about when I was a kid?

Are the lakes less polluted now than in the 1960s?

Has there ever been a massive oil spill on any of the Great Lakes?

Whatever happened to the Erie Canal?

Do the lakes ever completely freeze?

Which industry is most responsible for pollution of the Great Lakes?

Which is the most polluted tributary?

□ ■ □

This exercise was useful if it helped you inventory some of the things you're curious about. It was even more useful if you found something on your lists that raised interesting questions for you. You might have a tentative research topic. I found an item—the Great Lakes—in the list of "Things I know something about." I grew up a quarter mile from Lake Michigan, and it became a fixture in my life. It still is, though I now live 1,500 miles away. I wasn't quite sure what I wanted to know about the Great Lakes, but when I made my list of questions, my curiosity grew along with it. Any one of these "trail-head" questions might lead me into a great research project.

Other Ways to Find a Topic

If you're still stumped about a tentative topic for your paper, consider the following:

▪ *Search an index.* Wander over to the library and use InfoTrac, a CD-ROM index, or the computerized card catalog to hunt for topic ideas in broad subject areas that interest you. For example, type in PSYCHOLOGY, and you'll get a long list of "see also" prompts that will narrow the subject further. Choose one subject (e.g., forensic psychology) that sounds interesting to you, and follow the prompts. The topic will be further narrowed (e.g., forensic toxicology). Keep shaving away at the subject until you get a narrow topic that intrigues you.

▪ *Browse through an encyclopedia.* A general encyclopedia, like the *World Book, Encyclopaedia Britannica,* or *Encarta,* can be fertile ground for topic ideas. Start with a broad subject (e.g., the Great Lakes) and read the entry, looking for an interesting angle that appeals to you, or just browse through several volumes, alert to interesting subjects.

▪ *Surf the Net.* The Internet is like a crowded fair on the medieval village commons. It's filled with a range of characters—from the carnivalesque to the scholarly—all participating in a democratic exchange of ideas and information. There are promising research topics everywhere. Maybe begin with a site like *The Virtual Library* (http://www.w3.org/pub/DataSources/bySubject), which tries to organize Net resources by subject. Choose a subject that interests you, say autos or cognitive science, and follow any number of trails that lead from there into cyberspace.

▪ *Consider essays you've already written.* Could the topics of any of these essays be further developed as research topics? For example, Diane wrote a personal essay about how she found the funeral of a classmate alienating, especially the wake. Her essay asked what purpose such a ritual could serve, a question, she decided, that would best be answered by research. Other students wrote essays on the difficulty of living with a depressed brother or an alcoholic parent, topics that yielded wonderful research papers. A class assignment to read Ken Kesey's *One Flew Over the Cuckoo's Nest* also inspired Li to research the author.

- *Pay attention to what you've read recently.* What newspaper articles have sparked your curiosity and raised interesting questions? Rob, a hunter, encountered an article that reported the number of hunters was steadily declining in the United States. He wondered why. Karen read an account of a particularly violent professional hockey game. She decided to research the Boston Bruins, a team with a history of violent play, and examine how violence has affected the sport. Don't limit yourself to the newspaper. What else have you read recently—perhaps magazines or books—or seen on TV that has made you wonder?

- *Consider practical topics.* Perhaps some questions about your career choice might lead to a promising topic. Maybe you're thinking about teaching but wonder about current trends in teachers' salaries. One student, Anthony, was being recruited by a college to play basketball and researched the tactics coaches use to lure players. What he learned helped prepare him to make a good choice.

- *Think about issues, ideas, or materials you've encountered in other classes.* Have you come across anything that intrigued you, that you'd like to learn more about?

- *Look close to home.* An interesting research topic may be right under your nose. Does your hometown (or your campus community) suffer from a particular problem or have an intriguing history that would be worth exploring? Jackson, tired of dragging himself from his dorm room at 3:00 A.M. for fire alarms that always proved false, researched the readiness of the local fire department to respond to such calls. Ellen, whose grandfather worked in the aging woolen mills in her hometown, researched a crippling strike that took place there sixty years ago. Her grandfather was an obvious source for an interview.

- *Collaborate.* Work together in groups to come up with interesting topics. Try this idea: Organize the class into small groups of five. Give each group ten minutes to come up with specific questions about one general subject—for example, American families, recreation, media, race or gender, health, food, history of the local area, environment of the local area, education, and so forth. Post these questions on newsprint as each group comes up with them. Then rotate the groups so that each has a shot at generating questions for every subject. At the end of forty minutes, the class will have generated perhaps a hundred questions, some uninspired and some really interesting. Look for a question or topic you might like to research. Also look for opportunities to collaborate with others on a topic.

What Is a Good Topic?

Most writing—be it a personal essay, a poem, or an instruction sheet for a swing set—is trying to answer questions. That's especially true of a research paper. The challenge in choosing the right topic is to find one that raises questions to which you'd really like to learn the answers. Later, the challenge will be limiting the number of questions your paper tries to answer. For now, look for a topic that makes you at least a little hungry to learn more.

Also consider the intellectual challenge your topic poses and where you will be able to find information about it. You might really want to answer the question, Is Elvis dead? It may be a burning question for you. But you will likely find that you must rely on sources like the *National Enquirer* or a book written by a controversial author who's a regular guest on *Geraldo*. Neither source would carry much weight in a college paper. If you're considering several topic ideas, favor the one that might offer the most intellectual challenge and possibly the most information. At this point, you may not really know whether your tentative topic meets those criteria. The rest of this week, you'll take a preliminary look at some library sources to find out whether you have selected a workable topic.

Checking Out Your Tentative Topic

Consider the potential of the tentative topic you've chosen by using this checklist:

- Does it raise questions I'd love to learn the answers to? Does it raise a lot of them?
- Do I feel strongly about it? Do I already have some ideas about the topic that I'd like to explore?
- Can I find authoritative information to answer my questions? Does the topic offer the possibility of interviews? An informal survey? Internet research?
- Will it be an intellectual challenge? Will it force me to reflect on what *I* think?
- Are a lot of people researching this topic or a similar one? Will I struggle to find sources in the library because other students have them?

Don't worry if you can't answer yes to all of these questions or if you can't answer some at all just yet. Being genuinely curious about your topic is the most important consideration.

Making the Most of an Assigned Topic

If your instructor limits your choice of topics, then it might be a little harder to find one that piques your curiosity, but it will not be nearly as hard as it seems. It is possible to find an interesting angle on almost any subject, if you're open to the possibilities. If you're not convinced, try this exercise in class.

❑ *EXERCISE 1.2*
The Myth of the Boring Topic

On a piece of paper, write down the most boring topic you can think of. Anything. Exchange papers with someone else in class. Brainstorm a list of questions about your partner's proposed topic that might, in fact, be really interesting to explore. Play with as many angles as you can. Discuss in class what you came up with.

□ ■ □

Last time I tried this exercise, my partner's boring topic idea was dolls. I thought it was a brilliantly boring choice. But it didn't take me long to come up with some angles that were intriguing.

> Why do some dolls, like the Cabbage Patch, suddenly soar in popularity and then seem to fizzle?
>
> Why are some dolls, like Barbie and GI Joe, so enduring?
>
> How are new doll designs conceived?
>
> Why is my one-year-old daughter so bonded with the rattiest-looking doll in her collection?
>
> What's the nature of that bond?
>
> Would it be different for a one-year-old boy?

Suddenly, I was starting to getting excited about a topic I would have dismissed without a second thought.

If you've been asked to write on an assigned topic or choose from a limited list of ideas, don't despair. Instead, brainstorm a long list of questions, hunting for an angle you can get enthusiastic about.

If at first, you seem to have little interest or knowledge about an assigned topic, try bringing it into your world, as I did when I reflected on what I've noticed or experienced with dolls. Pay attention to what aspects of the topic get a rise out of you, if any. If you're asked to write a paper on an assigned novel, for example, which characters struck you? Which moments or scenes moved you? Why? Could your paper explore the answer to one of those questions?

Another way to discover what makes you curious about an assigned topic is simply to do some preliminary reading on it. Find some books and articles on the topic, and scan them for interesting angles or intriguing bits of information. What did you encounter that made you want to read more?

If all else fails, examine your assigned topic through the following "lenses." One might give you a view of your topic that seems interesting.

- *People.* Who has been influential in shaping the ideas in your topic area? Do any have views that are particularly intriguing to you? Could you profile that person and her contributions?
- *Trends.* What are recent developments in this topic? Are any significant? Why?
- *Controversies.* What do experts in the field argue about? What aspect of the topic seems to generate the most heat? Which is most interesting to you? Why?
- *Impact.* What about your topic currently has the most affect on the most people? What may in the future? How? Why?

Admittedly, it is harder to make an assigned topic your own. But you can still get curious if you approach the topic openly, willing to see the possibilities by finding the questions that bring it to life for you.

BEFRIENDING THE LIBRARY

I'm not exactly sure when I got my first library card, but I remember using it often. The children's section of my hometown library was a small wing, off to the side, with bright carpeting and miniature fiberglass tables and chairs in orange and green; even the bookshelves were pint sized. On long, boring summer days, I would wander into the library and through the shelves of books, pulling this one and then that, each covered in protective plastic. I would settle into one of those small chairs and spend an hour quietly turning pages. The librarians, always busy with official business behind the desk—stamping cards, flicking through files, taping those protective plastic covers on new books—would smile pleasantly at me now and then. I liked the children's library, and it seemed to like me.

The adult section, visible from the children's library through a glass door, was another matter. The shelves towered to the ceiling, and the tables and chairs were large, lumbering, wooden things that I could barely move. I was lost in the place. The books were out of reach, there were too many of them, and my sneakers squeaked on the bare, polished floor. The librarians, sentries of silence, seemed to watch my every move. The adult library seemed a solemn place, as if the business of containing so much knowledge was a serious affair.

Faced with spending the next five weeks in your college library, you may feel a little like I did as a child, wandering through that glass door to the adult section. Your college library may seem to be a wilderness, waiting to swallow you up. This apprehension is easy to understand. Some large universities have a dozen or more libraries on campus, each specializing in a field or discipline. The central library may contain vast collections of books, numbering in the millions. In addition to the familiar card catalog, there are computer terminals for accessing sources and a large section of specialized indexes and other reference materials. Everywhere you look, scholarly looking people sit before mounds of thick books, confidently flipping pages. A scene such as this may make you long for the days when writing a research paper involved buying a packet of 3" × 5" notecards and going directly to the *Encyclopaedia Britannica,* or maybe, if it was a serious research paper, the *Reader's Guide to Periodical Literature.*

When I poll my students every semester about their feelings toward the library, the result is always the same and almost unanimous: The college students I teach, freshmen and seniors alike, dread spending time in the library. Some are almost phobic about it. If you share some of these feelings about library work, then you're in for a rough time the next five weeks. Why do so many of us view the library with such loathing? It's worthwhile talking about that in class this week. But first, explore your own library experiences, as I did, in the following short exercise.

❑ *EXERCISE 1.3*
Loving or Loathing the Library

Take your research notebook, and go over to your library. Find a comfortable place to sit. Now spend seven minutes fastwriting about your experiences in libraries. Try beginning with your hometown library, where you spent time researching that paper on China in the seventh grade. Think about librarians you've known or where in the library you used to hang out. Maybe reflect on how you feel and what you think now, sitting in this library. Write about what excites you about doing research there and what you dread.

A reminder about fastwriting: Like brainstorming, fastwriting requires that you suspend judgment about what you think until you see what you say. Give yourself permission to write badly. You will be able to if you write quickly. Don't worry about grammar or even staying focused. Just keep your pen moving for seven minutes, thinking *through* writing rather than *before* it.

Discuss your fastwrite in class this week or with your instructor in conference.

□ ■ □

Inevitably, anyone who has been given a research assignment has had the experience of not finding a source that should have been there. In your fastwrite, you may have remembered researching a topic only to discover that the book you really needed was missing from the shelves or that the magazine with the essential article had key pages torn out of it. You may have written about the harsh glare of the fluorescent lights or how the heat was always turned up too high in your local library. You may have reminisced about carrying bundles of index cards for that paper on *The Scarlet Letter* and using one of them to pass love notes to Lori Jo Flink, only to discover later that the card contained information you needed.

But maybe you also wrote about the smell of new books, or the time you discovered the works of a favorite writer, or the kind librarian who first introduced you to *Charlotte's Web.* Maybe you wrote about some dark corner of the library where you pulled your chair up to a small table near the window, opened a good book, and lost yourself in another world some writer created.

For many of us, libraries that were once welcoming places during childhood become less so as we grow older. The disenchantment often begins in high school, when we are given research assignments that don't interest us or are asked to use sources that we don't know how to find. Pretty soon, the library becomes identified as the place where we toil over things we don't care about.

As I pointed out earlier, a successful research paper most often results from a good experience with the process of researching. Making it a good experience depends partly on finding a way to engage your topic, to make it your own, and to be motivated largely by your own curiosity. In essence, you need to get control over your topic by discovering the questions *you* most want answers to and by letting them guide your search. In the same way, you must learn to get control of the library and what it offers, rather than let the library keep control over you. Learn to see the library as an ally, rather than an adversary, in helping you to find out what you want to know.

Exercise 1.3 gave you a chance to air your feelings about research and libraries. This week, you also need to get down to the task of understanding how your library works, especially where to begin looking for the answers to your questions.

The Basic Plan of the College Library

Most university libraries are similarly organized. There's a *circulation desk* where you check out material, pay fines, ask for change for the photocopy machines, and so on. One valuable service the circulation desk provides the researcher is determining whether a needed book has been loaned and if so, when it's due back. The *stacks* is another name for a room full of bookshelves. A few stacks may be closed to students, which means the library staff will have to retrieve material for you, but most stacks are open, and you will be free to wander about. A *reserve desk* holds material, often at an instructor's request, for use by students for a few hours at a time. A room may also be set aside for *special collections,* or material that is valuable or unique.

The *reference section* is the research writer's most important resource. If research is like detective work, then the reference section is the FBI archives. The reference section is rich with helpful leads and information about what you want to know and where you can find it. Knowing just where to look will make an enormous difference in how you feel about the library. It may never be a place where you want to camp out, but once you've mastered the basics of the reference section, the library may finally become familiar territory.

Fortunately, there are people who can help. Most university libraries have staff who are specialists in reference materials, and there's often a separate reference desk where these librarians dwell, surrounded by computer terminals. Most librarians are experts at helping students with their research papers. For years, I was too shy to ask for help or afraid my questions would seem dumb. I now realize how much time I wasted, looking in the wrong places, and how much useful information I never found because of my reluctance to ask for help. Don't make the same mistake.

The Computer Revolution

Whether you love libraries or loathe them, by now, you likely know the basics of how to use them. You know about the general encyclopedias, the *Reader's Guide,* and the card catalog. You may even have tangled with microfilm or a computer database. But what you probably don't appreciate yet is that the reference room is an incredibly rich place to mine for information. And this resource keeps getting better as new reference materials are developed, along with new ways to access them. The computer revolution has transformed the reference room. In recent years, for example, my own university library has added many new computer stations and completely converted the card catalog to a new computer system that not only allows me to call up books by author, subject, and title but will tell me whether the book I want is available and if not, when it's due back.

CD-ROM

Compact disc (CD) technology, which has revolutionized the music industry, has done exactly the same thing for the library. The CD-ROM (compact disc with read-only memory) is now a standard fixture in most college libraries, and it's not difficult to use. Once you learn how to search for books and articles on your topic using the CD-ROM, you'll be hooked.

The typical CD-ROM is a 4.72 inch plastic disc that looks exactly like the latest issue from the Rolling Stones, but in some ways, it's even more impressive. Each disc can hold over 500 megabytes of information, the equivalent of 250,000 pages of text.

CDs run on standard personal computers and are available in many fields of study, from general science to business, indexing articles and books on virtually any topic. Using a CD, you can do a search in minutes that would take an hour or more using conventional, bound reference sources. For example, imagine searching ten years' worth of the *Reader's Guide* for articles on manic depression. You'd have to go through each year, volume by volume, jotting down promising article citations. The CD-ROM equivalent of the *Reader's Guide* (the most popular version is InfoTrac) would do the same search in about sixty seconds, and if it were linked to a printer, output the citations you select. You wouldn't have to pick up a pen.

There's a good chance that your college library is quite different than the library you began to master back home, especially in terms of size and sophistication. Before you plunge into researching your paper, get some practice in the reference room, using key sources and discovering new ones. The exercise that follows will introduce you to some of the basic references and how they can contribute to your paper. At the same time, you'll do some preliminary research on your tentative topic, which may help you narrow your focus. Later, you'll get some practice with more unusual sources that you may not have known existed; such sources can really help if you're having a hard time finding information.

❑ *EXERCISE 1.4*
Navigating the Reference Section

Navigate is the key word when it comes to working with the vast selection of reference materials. The reference room contains literally thousands of encyclopedias and indexes and almanacs and directories and databases and catalogs and bibliographies. As you make your way through this cluttered landscape, it's easy to get lost. Start with the basics, the reference sources most often consulted by writers of college research papers.

This exercise will give you a workout in the basics. It's a self-teaching exercise. You should work at your own pace; most students complete the work in about two to three hours. You'll need your research notebook to jot down the answers to the questions or the results of each step. If your instructor wants you to hand in the results of this exercise, you might want to do it on a separate piece of paper or perhaps he will give you a separate copy of the exercise to write on. Your instructor will also alert you to any special instructions regarding your own library.

If possible, use this exercise to begin researching your tentative research topic, if you have one. For some topics, however, that won't be possible, since this exercise is intended to introduce a range of reference materials, some of which may not be relevant to certain topics. But do every step anyway, even if you have to invent a search term. Try to get practice with each reference source.

Knowing What to Look For

Step 1: If you have a topic, describe it in a few words here.

____

____

____

____

____

____

Step 2: Now find the book entitled *Library of Congress Subject Headings (LCSH).* It may be available at the reference desk. Otherwise, ask the librarian where it's shelved, or check the card catalog. The *LCSH* lists the standard headings used by most libraries to catalog information. It's an easy way to find one or more headings that

Subject heading in boldface. The notation "(May Subd Geog)" indicates that a subject may also be subdivided according to geographic location (e.g., Animal Rights—United States).

UF *stands for "used for." It lists less suitable terms for the same subject.*

BT *means "broader term."* NT *means "narrower term."*

USE *is a code that lists the standard LC term under one that is not standard.*

Animal rights *(May Subd Geog)*
[*HV4701-HV4959*]
Here are entered works on the inherent rights attributed to animals. Works on the protection and treatment of animals are entered under Animal welfare.
UF Animal liberation
Animals' rights
Rights of animals
BT Animal welfare—Moral and ethical aspects
—Law and legislation
USE Animal welfare—Law and legislation
— **Religious aspects**
— — **Baptists,** [**Catholic Church, etc.**]
— — **Buddhism,** [**Christianity, etc.**]
Animal running
USE Animal locomotion
Animal sculptors *(May Subd Geog)*
UF Animaliers
BT Sculptors
Zoological artists

When subjects correspond to Library of Congress (LC) class numbers *(i.e., number classifications by subject areas), they are included here. These numbers can be very helpful if you just want to browse the shelves for books.* Scope notes *are sometimes added to explain headings.*

Subdivisions of the main subject heading, also in boldface.

FIGURE 1.1 There's no need to guess what subject headings to use when searching on your topic. The *Library of Congress Subject Headings* will get you off to the right start. Here a student looking for sources on animal liberation will discover that "Animal rights" is the heading to use.

will quickly yield useful sources on your subject (see Figure 1.1). Consult the *LCSH* with your topic in mind, and jot down several search terms that seem promising. In case you're considering skipping this step, see "The Story of a Search" at the end of this exercise (pages 56–59). It's a cautionary tale.

____

____

____

____

____

____

____

General Encyclopedias: Getting the Lay of the Land

I know you probably used encyclopedias extensively in high school. They're not nearly as useful for college papers, but they can be a good starting point to see the landscape of a subject, particularly if you're trying to narrow the focus.

The *Encyclopaedia Britannica* indexing system is in several sections: Micropaedia (ten volumes), Macropaedia (nineteen volumes), Propaedia, and Index. Some Micropaedia articles are complete in themselves, but others refer to one or more places in other Macropaedia volumes, which often expand on the subject. The Propaedia, among other things, explains how knowledge is organized in the encyclopedia and lists authors of articles in it.

Step 3: Find the *Encyclopaedia Britannica* index, and locate one or more listings that best match your topic. Look up the articles listed, and read each quickly. Select the article that seems most useful, and write down the title and something intriguing you gleaned from reading.

Title: ____

Intriguing thing: ______________________________________

____

Make sure to check the bibliography at the end of each encyclopedia article for other useful sources on your topic. If some citations seem promising, write them down below. Be certain to get all the bibliographic information you'll need, in case you need to document the source in your paper (author, title, publication information).

Promising citations in bibliography (if any): ________________

____

Surveying the Reference Landscape

Using a general encyclopedia is a little like looking through the wrong end of binoculars. You get the long view of a subject. That can be helpful, particularly if you are trying to further focus on some aspect of the topic. But in a college paper, pretty soon you'll find that the information in the encyclopedia does not provide the detailed glimpse at your topic that you need. The next step, consulting Balay's *Guide to Reference Book*s (in earlier editions, it was edited by Sheehy),* may

*Robert Balay, ed., *Guide to Reference Books,* 11th ed. (Chicago: American Library Association, 1996).

move you much closer to that detailed information by directing you to more specialized reference materials.

The *Guide* will reveal to you—maybe for the first time—the incredible variety of references that are available these days. In the *Guide,* organized by field of study (e.g., humanities, social and behavioral sciences, history, science and technology, etc.), you will find listings for specialized encyclopedias, almanacs, directories, indexes, handbooks, dictionaries, and bibliographies. Some of these sources—including almanacs, encyclopedias, and dictionaries—may directly provide information on your topic. But the most useful sources—such as bibliographies and indexes—contain citations for other books and articles that may have information you need.

Step 4: Find the *Guide to Reference Books* in the reference room of your library. Check the index in the back of the book, using the topic headings you found in the *LCSH* or any others that you think might work. Alternatively, look in the front of the book at the more general subject headings, and try to match one to your topic. After deciding what headings to use, browse through the appropriate sections in the *Guide* (see Figure 1.2). Note any reference sources (e.g., an index, almanac, bibliography, dictionary, encyclopedia, etc.) that you find in the *Guide* that seem promising, and list the title of one below. Make sure you get the complete bibliographic citation, and comment on what you hope to find in that source. (Don't bother to try to find any sources yet.)

Type of Source (circle): Index Almanac Bibliography

Dictionary Encyclopedia Directory Handbook Other

Author: ____________________

Title: ____________________

Publication information: ____________________

Comment: ____________________

____

FIGURE 1.2 A student researching violence on television in the *Guide to Reference Books* might check the index at the back of the book under "Television" and be referred to this page. Note the wide range of possible references that might be useful, especially the bibliographies. The student would then check to see which of the pertinent sources are in his library's collection.

Source: Reprinted with permission of ALA from *Guide to Reference Books,* 10th edition.

See also BG229–BG230, BG233.

Guides

Schreibman, Fay C. Broadcast television: a research guide. Ed. by Peter J. Bukalski. Los Angeles, Calif., American Film Inst., Education Services, 1983. 62p. (Factfile, no. 15) **BG290**

A useful guide for an area not well covered bibliographically.

Bibliography

McCavitt, William E. Radio and television: a selected, annotated bibliography. Metuchen, N.J., Scarecrow Pr., 1978. 229p., Suppl. one, 1977–81. Metuchen, N.J., 1982. 155p. **BG291**

For full information *see* CH503.

NAB broadcasting bibliography: a guide to the literature of radio & television. Comp. by the staff of the NAB Library and Information Center, Public Affairs Dept. 2d ed. Wash., Nat. Assoc. of Broadcasters, [1984]. 66p. **BG292**

1st ed. 1982.

Lists 360 books, most of them published since 1975, under seven categories (with numerous subdivisions): fundamentals of broadcasting, the business of broadcasting, broadcasting and the law, the technology and technique of broadcasting, broadcasting and society, comparitive broadcasting, related technologies. Also includes a list of periodicals and a publishers directory. Author/title index.

Dissertations

Kittross, John M. A bibliography of theses & dissertations in broadcasting, 1920–1973. Wash., Broadcast Education Assoc., 1978. [238]p. **BG293**

An author listing of some 4,300 dissertations and master's theses completed at American universities, with keyword-in-title index plus an index by year of completion and another by broad topics.

Sparks, Kenneth R. A bibliography of doctoral dissertations in television and radio. [3d ed.] Syracuse, N.Y., School of Journalism, Syracuse Univ., [1971]. 119p. **BG294**

A classified listing of some 900 dissertations completed through June 1970. Author index. Z7221.S65

Indexes

International index to television periodicals; an annotated guide. 1979/80– . London, Internat. Federation of Film Archives, [1983]– . Biennial. **BG295**

Michael Moulds, ed.

While Balay's *Guide to Reference Books* is perhaps the best guide of its kind, several alternatives are available. One I recommend is the following:

Walford, A. J. *Walford's Guide to Reference Material.* 6th ed. London: Library Association Publishers, 1993.

Finding Books

It may seem as if you've already put in your time thumbing through card catalogs, looking for books. You likely know the basics of this task. But because of the massive number of books in your college library and the new technologies for cataloging them, you could probably use some practice. First, some background on how books are classified and shelved.

There are two systems for classifying books: the *Dewey Decimal* and the *Library of Congress* systems. Each is quite different. The Dewey system, reportedly conceived in 1873 by an Amherst College undergraduate while daydreaming in church, is numerical, dividing all knowledge into ten broad areas and further subdividing each of these into one hundred additional classifications. Adding decimal points allows librarians to subdivide things even further. Just knowing the *call number* of a book will tell you its subject.

Here's the most basic division of Dewey numbers by subject:

000–099	General Works
100–199	Philosophy
200–299	Religion
300–399	Social Science
400–499	Language
500–599	Pure Science
600–699	Applied Science and Technology
700–799	Fine Arts
800–899	Literature
900–999	History

In the Dewey system, there's also a second number (or *Cutter number*) assigned to each book, which begins with the first letter of the author's last name. You don't need to know the significance of that number; just make sure you carefully copy it down when you're interested in the book.

The *Library of Congress* system, which uses both letters and numbers, is much more common in college libraries. Each call number begins with one or two letters, signifying a category of knowledge, which is followed by a whole number between 1 and 9,999. A decimal and one or more Cutter numbers sometimes follow. The Library of Congress system is pretty complex, but it's not hard to use. As you get deeper in your research, you'll begin to recognize call numbers that consistently yield useful books. It is sometimes helpful to simply browse those shelves for other possibilities.

Here is a list by subject of Library of Congress classifications:

A	General Works
B–BJ	Philosophy, Psychology
BL–BX	Religion
C	Auxiliary Sciences of History
D	History: General and Old World
E–F	History: America
G	Geography
H	Social Sciences
J	Political Science
K	Law
KD	Law of the United Kingdom and Ireland
KE	Law of Canada
KF	Law of the United States
L	Education
M	Music
N	Fine Arts
P–PA	General Philology and Linguistics; Classical Languages and Literatures
PB–PH	Modern European Languages
PG	Russian Literature
PJ–PM	Languages and Literatures of Asia, Africa, Oceania, American Indian; Artificial Languages
PN, PR, PS, PZ	General Literature: English and American
PQ	French, Italian, Spanish, and Portuguese Literatures
PT	German, Dutch, and Scandinavian Literatures
Q	Science
R	Medicine
S	Agriculture
T	Technology
U	Military Science
V	Naval Science
Z	Bibliography, Library Science, Reference

Your library may still have an active card catalog, and you probably already know how to use that. Remember that cards are organized by subject, author, and title. At this point, you'll likely look under promising subject headings, perhaps suggested by what you found in the *LCSH*. Make sure you pay attention to "see" and "see also" cards; they'll suggest other useful headings to check.

But it's likely your college library, like mine, has retired its 3" × 5" cards and replaced them with an *online card catalog*. This online system uses a computer to do the same thing that you used to do, thumbing through the card catalog. And of course, the computer is much faster. Search on a subject, author, or title, or on *keywords* that explain the topic you want to find. Figure 1.3 shows what a typical screen of an online catalog looks like.

Step 5: Give your online system (or card catalog) a try. Look up your tentative topic under one or several promising subject headings, and note at least one book that seems useful. (Be sure to list all the bibliographic information, in case you end up citing this source in your paper.) If your topic is current or doesn't otherwise lend itself to treatment in books, find a book on a subject that does interest you—

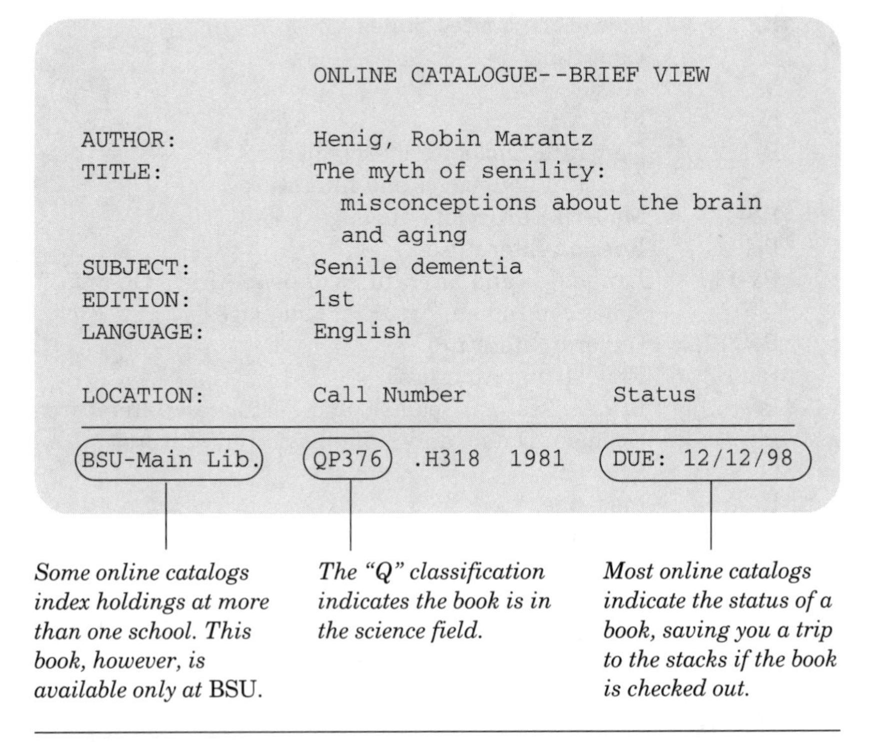

FIGURE 1.3 A typical screen of an online catalog. This one shows a brief display; a full display would include more bibliographic information on the book.

say, fiberglass repair. (To be sure, there aren't many topics that can't be researched in books.)

Call number: ______________________________

Author(s): ______________________________

Title: ______________________________

Place of publication: ______________________________

Publisher: ______________________________

Date of publication: ______________________________

Now retrieve the book from the stacks. Look for a wallchart that lists on what floor each category of call numbers is shelved, or ask the reference librarian for help. At a larger university, you may discover that the book you need is in another library on campus. For example, my college has separate physics, engineering, bioscience, and nursing libraries. Getting a book from one of them involves a short hike.

If the book is missing, check to see if it was misshelved or may be on the "waiting shelves," which are where books that have recently been returned are held before they're reshelved by the library staff. Finally, consult someone at the circulation desk to find out if he knows where the book might be.

If you have found your book, write down the first sentence of the first full paragraph on page 10:

Sentence on page 10: ______________________________

____

If you haven't found your book, explain what you discovered about its fate at the circulation desk (when it's due back, whether it's being rebound, if it's lost, etc.).

Status of the missing book: ______________________________

____

Interlibrary Loan

If the library doesn't have the book (or journal article) you want, don't despair. If you have enough lead time (a few weeks), the library can get the book from another library through interlibrary loan, a service provided at most college libraries through the reference desk. It's usually free and simply involves filling out a search form.

Checking Bibliographies

At the backs of many books are bibliographies, or lists of sources that contributed to the work. Sometimes bibliographies can be mined for additional sources that might be useful to you.

If the book you found has a bibliography, scan the titles and write down at least one promising source you find.

Source(s) from Bibliography: ______________________________

____

____

Finding Magazine and Journal Articles

It used to be that the table where the *Reader's Guide to Periodical Literature* was located was one of the busiest in the reference room. You likely remember the *Reader's Guide* from high school—it and the encyclopedia were often the major reference sources for almost every paper. The bound version of the *Reader's Guide* is still useful, especially if you're hunting for popular magazine articles published before 1990, which may not be included in the new computer indexes. The *Reader's Guide* is also helpful for locating citations of magazines published last month, which are listed in monthly supplements.

Perhaps the real weakness of the *Reader's Guide* for college papers is that it's mostly an index to nonscholarly magazines, such as *Time, Redbook,* and *Sports Illustrated.* There's nothing wrong with these publications. In fact, you may end up using some in your paper. But as you dig more deeply into your subject, you'll find that the information in popular magazines will often begin to tell you what you already know.

Fortunately, the *Reader's Guide* is now available in electronic format, which makes it easier to use. Other electronic databases that index popular periodicals include *Magazine Index* and *Periodical Abstracts.* But perhaps the most useful of all is a computerized index called *InfoTrac,* which includes over two thousand periodicals in a variety of disciplines. InfoTrac includes some scholarly journals that contain articles written by experts in their fields. It also indexes the *Wall Street Journal* (the last six months) and the *New York Times* (the last sixty issues). InfoTrac listings are updated monthly and cover the current year and the three years before that. If you're looking for periodical sources from before 1990, use the *Reader's Guide,* which began indexing general periodicals in 1890.

Using InfoTrac is easy. You can search by subject, author, title, or keyword and in most cases even get a printout of the sources you find. One of the nice things about using a computerized index such as this is that it will often help you narrow your search with the "see also" prompts.

Step 6: If your library has InfoTrac (or some other CD-ROM index to popular periodicals), give it a spin. Review the commands, and begin a search for articles on your topic. Of course, ask for help if you need it. Find a useful citation, and record the following information about it. (You might also want to attach the printout you received and hand it in with this exercise.)

Subject heading: ______________________________

Author: ______________________________

Title: ______________________________

Title of periodical in full: ______________________________

____

Volume or issue number and date: ______________________________

Pages covered by the article: ______________________________

If your library doesn't feature a computer index to periodicals, check the *Magazine Index* or *Reader's Guide* for sources, starting with the most recent issue (unless coverage of your topic is limited to a specific time).

If using InfoTrac didn't produce many sources for you, *try different subject headings.* This is really important. (See "The Story of a Search" at the end of this exercise.) Also consider if perhaps articles on your topic were published before 1990. Check the printed version of the *Reader's Guide* for older articles, but remember that one measure of the value of a source in an academic paper is its currency (depending on the subject, of course). Finally, it's possible, though unlikely, that your topic just doesn't lend itself to treatment in periodicals. If so, try another topic that interests you to get some practice with periodical indexes.

You may discover, much to your delight, that InfoTrac provides you with a long list of periodicals that seem helpful. But there may be a hitch: Your library may not have all of them. You may be able to find that out quickly by consulting a catalog produced by your library that lists its periodical subscriptions and where each is located. The catalog is often a bound computer printout, frequently located near the periodical indexes. Ask the reference staff where this list is.

Does the library have the periodical you listed above? ___ *yes* ___ *no*

If so, where is it located? ______________________________

____

For the moment, don't bother hunting down any periodicals. You'll have time for that later. Generally, periodicals published more than a year ago are bound together and shelved alphabetically on a floor in the main library or in one of the satellite libraries on campus. Current issues, published within the last year, are often shelved separately, unbound.

Indexes to Specialized Periodicals

Although *Time* and *Sports Illustrated* were great sources for that paper on steroid use by high school athletes you wrote for English your senior year, you will soon find that as you research this college paper, popular periodicals will quickly stop telling you much that's new to you. Popular periodicals can be good sources for memorable quotes, anecdotes, or case studies, but often, the articles are written by nonexperts whose treatments of topics are fairly superficial, at least for academic papers. You need to dig deeper.

If you used InfoTrac for Step 6, you may have discovered a few periodicals appearing on your printout—such as *Science, Foreign Affairs,* and *Psychology Today*—that are somewhat more authoritative than everyday magazines, partly because they are written for a less general audience. They are still popular periodicals, however, and of limited use.

The college researcher should make a practice of consulting scholarly publications, whenever possible. Simply put, these periodicals are written for and by people in their respective fields. For example, the *American Journal of Political Science* is one of a handful of professional periodicals read by political scientists. English instructors might read *College English*, and psychologists, *Psychology Review*.

Getting through articles in scholarly journals like these may sometimes be difficult—the terminology may be unfamiliar to you, and the prose, pretty dense. But often, the task is well worth the effort because you'll uncover information on your topic you won't find anywhere else. With a little practice, you'll learn to skim journal articles for useful information.

Sometimes journal articles are reports of new studies, which makes them *primary sources,* or feature analyses by people who are leaders in their fields. (See "Primary over Secondary Sources" in Chapter 2 for further explanation.) Skillful use of scholarly sources in your paper can be a big boost. They not only enhance the authority of your paper, but you'll discover that through your familiarity with some of the important thinkers on your topic, you will become something of an expert yourself.

But how do you find scholarly sources? Not surprisingly, there are computer databases (many now on CD-ROM) and bound indexes to choose from. If your library has CD technology, use it; it has enormous advantages in searching for journal articles. But you should be familiar with the key bound indexes, too. They're listed below. All are published by the same company that produces the *Reader's Guide,* so they're similarly organized and used.

SIX KEY JOURNAL INDEXES

Humanities Index (1974–date)*
Covers roughly 260 journals in archaeology, classical studies, language and literature, area studies, folklore, history, performing arts, philosophy, religion, and theology.

Social Science Index (1974–date)*
Covers 263 journals in anthropology, economics, environmental science, geography, law and criminology, medical science, political science, psychology, and sociology.

General Science Index (1978–date)
Covers about 115 journals in astronomy, atmospheric science, biology, botany, chemistry, environment and conservation, food and nutrition, genetics, mathematics, medicine and health, microbiology, oceanography, physics, physiology, and zoology.

Art Index (1929–date)
Covers about 300 English and foreign language journals dealing with subjects like art, archeology, architecture, design, and city planning.

Education Index (1929–date)
Indexes well over 300 journals in the arts, audiovisual education, comparative and international education, computers in educa-

*If you want to search for sources prior to 1974, check the predecessors to this index: the *International Index* (1907–1965) and the *Social Science and Humanities Index* (1965–1974).

tion, English language arts, health and physical education, language and linguistics, library and information science, multicultural/-ethnic education, psychology and mental health, religious education, science and mathematics, social studies, special education and rehabilitation, and education research.

Business Periodicals Index (1958–date)
Covers journals in the following subjects: accounting, advertising and marketing, agriculture, banking, building, chemical industry, communications, computer technology and applications, drug and cosmetic industries, economics, electronics, finance and investments, industrial relations insurance, international business, management, personnel administration, occupational health and safety, paper and pulp industries, petroleum and gas industries, printing and publishing, public relations, public utilities, real estate regulation of industry, retailing, taxation, and transportation.

If your library has computer workstations set up to use CDs, then you may rarely have to consult printed indexes to find journal articles on your topic. The books *are* worth checking when the database you consult doesn't cover the years you need or you want to look for articles published in the last few months. Though CDs are usually updated several times a year, they frequently don't contain the most recent issues of the journals they cover.

The indexes listed above are also *general indexes* of academic journals. There are a multitude of printed *specialized indexes* you might want to check that aren't on CD. For example, if you're writing a paper on deforestation in the Pacific Northwest, it might be worth checking the *Environmental Index* as well as the *General Science Index* on CD.

But in most college libraries, the new CDs are the easiest, most efficient way to search for academic articles. Between fifteen hundred and three thousand databases are available to libraries on CD, and more appear everyday. If you're unfamiliar with this medium, see the box "Tips for Searching on a CD." Many of the general indexes listed above are also on CD. The following is a list of additional databases that are popular with college researchers:

ABI/INFORM covers business, economics, and management topics, as well as the health care industry. It indexes about nine hundred journals in those fields, a third of which include not just citations but the full texts of the articles.

TIPS FOR SEARCHING ON A CD

Despite the explosion of databases on CD in a broad range of fields, they are often quite similar to use. Here are a few general points that should help you search successfully on most CDs:

1. *Computers take your mistakes seriously.* If you misspell a word, the computer will take you literally and search for it, as is. Ask the computer to search a database for articles on "Acholoics Anonymous," and you'll come up empty handed. Check your spelling, and you'll save time.
2. *Search topics need not be described in full sentences.* Actually, it's better to break your topic down into several headings that seem useful. If you're looking for information on the effects of medication on hyperactive children, break the topic down into three concepts—hyperactivity, medication, and children—and then search for articles about each. Most CD databases come with books of indexing terms, or subject headings—a sort of *Library of Congress Subject Headings* equivalent for computers. These manuals are very useful for finding search terms on your topic.
3. *Understand the difference between* ***and*** *and* ***or.*** These are not words the computer will search for, but they offer it key instructions about what to do. The word *or* tells the computer to look for synonyms, broadening your search. For example, if you ask it to search for "food *or* nutrition *or* diet," the system will retrieve records with any of these terms. In this case, it would produce a downpour of articles. The word *and* narrows the search, telling the computer to locate only those records that contain the words you specify. For example, telling the computer to find "hyperactivity *and* medication *and* children" will only produce records containing all those terms. (For tips on Internet search terms, see "Query, Query, Quite Contrary," pages 67–69).

ERIC indexes education-related periodicals. It's the electronic equivalent of the *Current Index to Journals in Education* and *Resources in Education.* ERIC is an enormously useful database for a variety of subjects, even those that are not exclusively education related.

Essay and General Literature indexes articles in the humanities and social sciences that are parts of edited collections.

MLA Bibliography is the key index for literature, linguistics, languages, and folklore, covering about three thousand journals in those fields. The CD version covers nine years on one disc.

Medline is the computerized version of *Index Medicus*, a widely used reference in the fields of medicine, pharmacy, pharmacology, and nursing.

PAIS (or the Public Affairs Information Service) covers politics, government, economics, and international and consumer affairs. It is also available in a bound version.

Psyclit is the computerized version of *Psychological Abstracts*. It indexes about fourteen hundred journals and is one of the most widely used databases in the field of psychology.

Sociofile indexes over fifteen hundred journals in sociology, anthropology, and social work and also covers some areas of education, health, and psychology.

Step 7: Search one or more of the indexes above for scholarly articles on your topic. Use either the bound indexes or, if your library has them, the appropriate databases on CD. Try several subject headings.

For *one* citation that seems promising, take down the following:

Subject heading: ______________________________

Author(s): ______________________________

Title of article: ______________________________

Title of periodical: ______________________________

Volume, date, page numbers: ______________________________

Does the library have it? ___ *yes* ___ *no*

If you came up empty handed, you may have to find a more specialized index in the field you're working. You'll have the chance to do that in the third week (see "Second-Level Searching" in Chapter 3). It's also possible that your topic simply hasn't been of scholarly concern. That doesn't necessarily mean you should abandon your idea. You may be able to find plenty of good sources other than journal articles. But for practice, use one of the indexes and look up something else that interests you, such as alcohol abuse patterns among college students.

If you did come up with some promising journal citations, don't bother hunting them down now.

Newspaper Articles

If your tentative research topic is local, current, or controversial, then newspapers can be useful sources. You'll rarely get much in-depth information or analysis from newspapers, but they can often provide good quotes, anecdotes, and case studies as well as the most current printed information available on your topic. Newspapers are also sometimes considered primary sources because they provide first-hand accounts of things that have happened.

Some newspapers are indexed. Generally, only those publications considered *national newspapers* are catalogued and saved, usually on microfilm but sometimes on CD. Among the national publications are the *New York Times,* the *Los Angeles Times,* the *Christian Science Monitor,* the *Washington Post,* and the *Wall Street Journal.* The index for each national newspaper is a separate bound volume or may be included on a computer database, such as Newsbank, or a microfilm index, such as the Newspaper Index.

Your college library probably subscribes to a variety of local or state papers, as well, which are often not indexed. Some may be on microfilm going back many years. If your topic involves an event that occurred at a specific time—say November 22, 1963—then you can simply scan the unindexed newspapers for coverage on and around that date.

Step 8: Check the index for the *New York Times* (or any of the other bound volumes), a computer database like Newsbank, or the *Newspaper Index* for articles on your topic. Try several subject headings, and note the "see also" prompts. If your topic doesn't lend itself to coverage in newspapers, find microfilm of the *New York Times* edition that was published on the day of your birth. If your library has a machine for photocopying from microfilm, make a copy of an article on your topic or the front page of the *Times* published on your birthday. Attach the photocopy to this exercise when you hand it in. If there is no copying machine, take down the following information:

Subject heading: ______________________________

Title of article: ______________________________

Newspaper: ______________________________

Date and page number: ______________________________

Government Documents

The United States government is the largest publisher in the world, and if your college library is a *depository* for government documents, it may receive almost everything the federal government publishes. Usually, there's one depository library in each state. Nondepository libraries may still receive government documents, though, and most college libraries do have a selective collection that is sometimes catalogued separately.

The great thing about government documents is that they cover a broad range of subjects. The bad news is that you may go nuts trying to find what you're looking for. Because government publications often arrive daily at your campus library, in one big pile, the staff has its hands full, cataloguing and shelving the material. Government document librarians sometimes live on the edge of chaos.

As always, there are printed and computer indexes to government documents. The most important index, the *Monthly Catalog of*

Indians of North America — Census, 1990.
American Indians and Alaska Natives—the Census counts for you! (C 3.272:D-3207 (I/AK)), 90-9318

Subject title

Indians of North America — Child welfare.
Foster care : use of funds for youths placed in the Rite of Passage program : briefing report to the Honorable George Miller, House of Representatives / (GA 1.13:HRD-87-23 BR), 90-8009

Title of document; brief description of type (report, study, hearing documents, etc.)

Regulating Indian child protection and preventing child abuse on Indian reservations : report (to accompany S. 1783). United States. Congress. Senate. Select Committee on Indian Affairs. (Y 1.1/5:101-203), 90-5566

Entry number begins with year of publication; keyed to separate volume with fuller description of document

Indians of North America — Children — Crimes against.
Federal government's relationship with American Indians : hearings before the

FIGURE 1.4 Using the *Monthly Catalog of Government Publications* is usually a two-step process. Check the subject/title/keyword index for pertinent headings. Then write down the *entry number* for a promising document, and look for the number in the separate volumes of entry numbers, which are listed sequentially. In this figure, the researcher is looking for a document on child abuse on American Indian reservations.

Entry number

SuDoc number (call number to help locate document in stacks)

Physical description of document; black dot indicates routinely sent to depository libraries

90-5566

Y 1.1/5:101-203

United States. Congress. Senate. Select Committee on Indian Affairs.

Regulating Indian child protection and preventing child abuse on Indian reservations : report (to accompany S. 1783). — [Washington, D.C.? : U.S. G. P. O., 1989]

12 p. ; 24 cm. — (Report / 101st Congress, 1st session, Senate ; 101-203) Caption title. Distributed to some depository libraries in microfiche. Shipping list no.: 89-803-P. "November 13 ... 1989." ●Item 1008-C, 1008-D (MF)

1. Child abuse — Law and legislation — United states. 2. Indians of North America — Child Welfare. 3. Indians of North America — Children — Legal status, laws, etc. I. Title. II. Series: United States. Congress. Senate. Report ; 101-203. OCLC 20777673

FIGURE 1.5 Here's what the researcher discovered using the entry number for the document on American Indian child abuse.

United States Government Publications, is available in both forms (see Figures 1.4 and 1.5). You can search a document in the *Monthly Catalog* by author, title, subject, series, stock number, and title keyword. Your library may have the *Government Publications Index,* which is drawn from the *Monthly Catalog,* available on CD.

Step 9: Check one of the indexes mentioned to locate any government documents on your topic that may be useful. Try a few possible subject headings, and go back a few years if you need to. If you find something interesting, take down the bibliographic information below.

Subject heading: ____________________

Full name of government branch that published document: ____________________

____

Title: ____________________

Date: ____________________

*Place and publisher:** ____________________

*Usually, "Washington, DC: Government Printing Office."

If you got nowhere looking in a government publications index, it's possible that your topic just doesn't lend itself to this type of source. The best bets, obviously, are topics that have to do with law, government, public policy, and the like. If you don't find anything on your topic, just for practice, try a topic that will be indexed—say, student loans.

If you did uncover something promising and your library doesn't have it, you can often send for the material by writing to:

Superintendent of Documents
Government Printing Office
Washington, DC 20402

It usually doesn't cost anything to get materials, and they should arrive in a few weeks, which is just enough time to be useful for your paper.

Don't bother locating the document in your library. You can do that later. For now, reward yourself with an Almond Joy. You've finished the library exercise and are familiar with the most important reference sources used by college researchers. You also may be well on your way through the initial stage of research on your tentative topic. In the second week, you'll refine your research strategy and learn about a few more reference sources that are specialized. These can be especially useful if you're having trouble finding information.

□ ■ □

The Story of a Search

You've finished your long journey through the reference section. Now, a parable.

Tim read an article in the local paper about a study on the academic performances of girls and boys in elementary school. It showed, among other things, that boys did significantly better than girls in math and science and speculated that perceived gender roles had something to do with these differences. Might some teachers call on boys more frequently than girls in a math class? Might some girls believe that they're not supposed to be good at science?

Tim wondered, too. He told me the focusing question for his research essay was Do boys and girls have different learning styles? I thought it was a good topic—a little broad maybe, but a good start. I was certain he'd find lots of information, *if* he could refine his search terms.

The next time I saw Tim, he had switched topics. I asked him why.

"Couldn't find anything," he said. "Nada."

What did he look for exactly, I asked.

"I looked for stuff on gender and on education," he said. "I found a ton of books and articles but nothing on gender and academic performance."

At first, I couldn't convince Tim to try again with better search terms—he was already on to another topic. And without checking the *Library of Congress Subject Headings* book, I didn't have any suggestions for new terms. I couldn't help feeling that he had surrendered too soon on a promising topic.

Based on the experience that Tim and countless other students have had, I've come to believe more strongly than ever that *the words you choose for your search can make or break your project.* Not using the right keywords, in the right combination, for a library or Internet search might mean you will find absolutely nothing on your topic when, in fact, there's a lot out there. Or using the wrong keywords might mean being deluged with information that requires you to spend hours sifting through the debris to find what you need.

So what are the right search terms? For library research, they're the terms about your topic included in the *Library of Congress Subject Headings (LCSH),* used in the proper combination so your search is sufficiently narrowed. (For tips on Internet search terms, see "Query, Query, Quite Contrary" later in this chapter.)

Both online card catalogs and CD-ROM indexes take the Library of Congress's lead on subject headings, so you simply have to check the *LCSH* book for ideas on what terms to use for your topic. CD-ROM databases usually include thesauruses that are based on the *LCSH;* you can use them, too. But online card catalogs and CD-ROMs also have features that can help you combine terms in useful ways. They use something called *Boolean operators*—AND, OR, and NOT—that tell the computer exactly what you have in mind. If you put an AND between search terms, you're saying that all of the terms must be in the title or abstract of the document. An OR between terms means any *one* term can be present. A NOT before a term means it should be excluded (e.g., air AND pollution AND Washington NOT DC).

I decided to issue Tim a challenge: If I could find good sources on his topic—the learning styles of boys and girls—might he reconsider his plan to abandon it? I started with the *Library of Congress Subject Headings,* looking under "Learning, Psychology of." Several narrower terms under that broad heading seemed promising: "Motivation in education" and "Learning ability." I tried them both using the online card catalog, just to see what kind of books would turn up. "Learning ability" produced twenty-five possibilities, including one promising text titled *Educability and Group Differences*. "Motivation in education" produced sixty-seven books, which was too many to sift through. But the *LCSH* suggested the narrower term "Academic achievement," which I then tried. That produced forty-seven books, including *Personality and Educational Achievement,* a text that would likely discuss gender and learning styles.

Next I moved to *Psychlit,* a CD-ROM index to psychology-related journals, and there, I really hit the jackpot. As you look at the following list, notice the sequence and combinations of search terms I tried and how I subsequently narrowed the number and potential relevance of the resulting citations.

1. *Search terms:* "academic achievement"
   Result: 3,470 citations
   Note: Ugh. Too many.
2. *Search terms:* "academic achievement **and** gender"
   Result: 273 citations
   Note: Still too many to sift through. Since Tim wanted to focus on boys and girls, I decided to narrow my next search to studies of elementary school students.
3. *Search terms:* "academic achievement **and** gender **and** elementary"
   Result: 73 citations
   Note: Better. But since the newspaper article that inspired Tim's topic focused on the differing performances of boys and girls in math, I decided to narrow the search to that subject.
4. *Search terms:* "academic achievement **and** gender **and** elementary **and** math"
   Result: 11 citations
   Note: This number of articles is clearly manageable; even so, I decided to play with the search terms one last time. I decided to try two of the terms suggested by the *LCSH* using the OR operator. Maybe one or the other—in combination with gender, elementary, and math—might produce some new articles. It did.
5. *Search terms:* "(academic achievement OR learning ability) **and** gender **and** elementary **and** math"
   Result: 11 citations
   Note: This search was wonderfully productive. *All* eleven articles were relevant to Tim's project—for example, "Gender differences in academic achievement: A closer look at mathematics" and "Gender differences in general academic self-esteem."

This story of a search ends well. Tim wrote a perceptive essay contrasting how boys and girls perform in elementary school. And like any good parable, this story has a moral: *Choose your words carefully.* The success of library searches often depends on how cleverly you choose and manipulate your search terms. For that reason, the *Library of Congress Subject Headings* may be the most important book in the library.

BEFRIENDING THE INTERNET

The Internet is an information resource that acts like I did when I was thirteen: at times, forthcoming, amiable, and responsible and at other times, rebellious, disorganized, and unreliable. Despite this schizophrenic character—or perhaps because of it—the Net can inspire among its users an obsessive relationship that rivals anything you experienced in the eighth grade: The Internet can drive you crazy, yet you can't seem to get enough of it. I have frequently seen my students spend hours browsing the Net for information, wandering from one end of cyberspace to the other, when they could walk across the quad to the library and find what they needed in less than thirty minutes.

Three Drawbacks of Internet Research

For all the its promise as an information source for the researcher—and the Internet's potential is really quite stunning—know that, for now at least, its drawbacks are worth considering. Let me mention a few:

1. *Information on the Internet is disorganized.* While search tools like Veronica, WebCrawler, and Yahoo and indexes like the Virtual Library are beginning to fence off some of cyberspace and organize some of the information that rushes helter-skelter into that virtual village commons every day, there is no limited set of comprehensive reference sources. This means you cannot be certain that a single search, using one of the many so-called search engines like Yahoo, will ever offer adequate coverage on what might be available on your topic. To research effectively on the Internet, then, you have to be extra resourceful. You have to learn how to launch multiple searches that will explore as much of that commons as possible.

2. *Information on the Internet is unreliable.* The free-for-all atmosphere of the Internet makes it a surprisingly democratic forum for ideas, debate, and dialogue. As a participant, I find that enormously

appealing, but as a researcher, the openness of the Internet makes me nervous. How reliable is the information I find there? Will it still be there tomorrow? And how can I establish the authority of Internet sources, especially since so much information is authored anonymously? For the moment, the Internet is not nearly as good a source for scholarly information as the library. Peer-reviewed journals—publications that will only print an article after it has been scrutinized by other experts in the discipline—are the most authoritative sources for college research papers, and very few of these journals are available online, though more appear every day. Moreover, some information on the Internet is downright wacky. So, while there *is* useful material to be found online, the researcher must, as Ernest Hemingway put it somewhat more graphically, always keep her "crap detector" on.

3. *It's easy to go nowhere very slowly.* Navigating the Internet can be like driving in Boston: You're constantly confronted by no street signs and complex intersections that seem to end up in rotaries. In fact, moving around in cyberspace, looking for information, is a lot like that: tentatively trying one street, then returning to try another—moving forward, then back, then forward, then back—until you find your way to a useful destination. Then, there are the breakdowns. *Servers*—the software programs that make information available to other computers on the Net—are sometimes fickle things. They can crash at the most inconvenient times.

Internet forays require patience, so if you're easily discouraged, online research may drive you batty. To avoid that, do what I did in Boston: First, learn to drive to Fenway Park without getting lost. There are a number of Internet sites that are fairly easy to find and frequently rewarding sources for college researchers; many of them are listed in this book. Learn to find and use these sites first; then, when you're more confident, you can try to find the Internet equivalent of the underground parking garage at the Boston Commons.

Three Reasons to Use the Internet for Research

Now that I've sobered you up about the drawbacks of online research, consider its unique advantages:

1. *The Internet is a marvelous source for topical information.* Do you want to read the March 23 *New York Times* article on the relationship between writing and Alzheimer's? Check the *New York Times Online.* Want to read the text of the president's most recent speech on funding student loans for higher education? Search *Fed-*

world's Web site for an FTP (file transfer protocol—see page 62) file of Clinton's speeches. If it's current information you're after, the Internet offers fast access to it through online newspaper and periodicals, commercial information services, and government sources. And the Internet is a particularly good resource for timely information on things to do with computer technology and the Net.

2. *The Internet is most useful when you're after something specific.* Say you want some information on how the population of Bosnia-Herzogovinia breaks down according to religious affiliation. Or perhaps you need an electronic version of Hawthorne's *The Scarlet Letter*, or you're interested in getting more information from Alcoholics Anonymous (AA). You will likely find all of this kind of specific information much more quickly on the Net than at the library. If you're not exactly sure what you're looking for or have only a broad notion of what it might be, the Internet can be a swamp. That's why so-called keyword searches, if they're carefully expressed, can produce useful information much more quickly than broader subject searches. (As you'll see, working your way through a subject search *can* be quite useful. But you can't be in a hurry.)

3. *Online documents can be easy to transport.* Downloading files or printing documents from the Net is easy, particularly with the new browsers, like Netscape. And once downloaded into your computer or onto a floppy disk, these documents are in an electronic format, which (if you write on a computer) makes it easy to move the information into your own document. Suppose you need a passage from *Huck Finn,* or a quote from President Clinton's latest speech on health care, or a table on drug use among high school students. Using the "Edit" and "Move" or "Cut" and "Paste" features of your word-processing program, you can drop the passages or tables into your paper (assuming the texts are compatible or can be converted by your program). However, the ease of transporting online documents should make you extra vigilant about avoiding plagiarism. Carefully cite and attribute any material in your own paper that is not your own.

A Cluttered and Colorful Canvas

At the risk of cluttering our discussion of cyberspace with one more metaphor, I'd like you to imagine a bad imitation of Jackson Pollock, the paint-splatter artist. Since the 1960s, when the Internet was first developed by the U.S. military, people and institutions have been throwing wet paint at what was once the large blank canvas of cyberspace, creating runny, amoebalike blobs of color that represent separate, though often overlapping, information resources. In other words, over the years the Net's creators have made more and more information available in what seems an almost haphazard, chaotic way.

Fairly early on, for example, someone heaved a bucket of color that represents *FTP space*, or file transfer protocol. This is both a source of files on a range of subjects located throughout the world on remote computers *and* a tool for getting these files. Something called *Gopher space* was another early paint blob that has continued to grow, applied energetically by researchers at the University of Minnesota, where Gopher originated. There are Gopher sites around the world on countless subjects, into which you can burrow if you know what you're looking for. And if you don't, there is a search tool called *Veronica* that can help you.

Other paint blobs include *Telnet, Usenet, e-mail,* and *WAIS* (or wide area information server). Each offers some unique resources. *Telnet,* for example, is a tool for accessing university libraries around the world, among other things. *WAIS* provides large databases on a range of subjects. *Usenet* features discussion groups among individuals who share an interest in topics from Disney movies to disarmament.

Now imagine that on top of this mess of color we throw two cans of bright paint that represent the *World Wide Web (WWW).* Runny fingers reach into almost every other color on the cyberspace canvas, overlapping with FTP space, Gopher space, and all the others. Now we have our masterpiece, except that the paint never dries on this one: Blobs of paint grow and continue to overlap.

While this chaos of color may seem impenetrable to the novice, the veteran Internet researcher learns to understand the meaning of each color and how she can reach it. Thankfully, the Web—that large splash of color in the middle of the canvas—offers beginners fairly easy access to many parts of the painting, Thus, the Web will be our vehicle for the quick tour that follows.

The Tangled Web

The World Wide Web began in 1989, when some European researchers saw the need for a tool that could access the range of documents, databases, and media on the Internet using a consistent interface. In the years that followed, a number of so-called browsers were developed for this purpose, which often used graphics instead of text-based commands to search for and retrieve information. The most famous of these is called *Mosaic,* but in the past several years, Netscape has assumed a dominant position as the most popular WWW browser on college campuses. So I'll often use Netscape as an example.

The stunning growth of the World Wide Web in the last few years is partly due to how easy it is to navigate cyberspace using browsers like Netscape and also the ease of moving between documents and Internet sites using something called *hypertext*. Most of the Web documents you can retrieve are sprinkled with terms, phrases, and titles that are highlighted in a different color from the rest of the text. Each piece of highlighted text—or *hotspot*—provides a *link* to other documents, media, and locations on the Internet that are related to the original document's subject. And best of all, each link is only a mouse-click away. Essentially, then, you can visit a Web site and from that one location wander in multiple directions to explore other leads that may seem promising. From a single Web site, you can reach into the edges of the multicolored canvas of cyberspace, exploring not just other Web sites but Gopher space, WAIS space, FTP space, and others.

Using a Browser

Before you journey into cyberspace, you should know the vehicle that you'll be driving. The WWW browser is your navigation tool, and while it's fairly easy to use, you should know how to do several key things before you start. The great majority of campus networks use Netscape (see Figure 1.6), but if you're working from home, you probably have access to a browser from a commercial network like America Online or Prodigy or from some local provider. These servers don't differ too much, and they have several key features in common. Use the "Help" feature on your browser and familiarize yourself with how it works, but pay special attention to the following:

- *The "Hotlist" or "Bookmark."* Do you want to have a really bad day? Try to find your way back to a Web site that was really important to your research after not writing down the Internet address of the document or not storing it in your "Hotlist." Browsers all provide easy ways to save the addresses of useful sites during your visit through such "Hotlists" or "Bookmarks." The feature usually involves simply clicking on the "Hotlist" or "Bookmark" button and agreeing to save the document title and Internet address on your personal list of important Internet locations. This is one of the Web browser's most valuable features. Use it regularly. If you're using a school computer, however, you may not be able to use the "Bookmark" function for personal information. In that case, carefully write down the addresses of important Web sites. Fortunately, most browsers also automatically record where you've been during a given session. This is a feature that you can find under the "Window"/"Window History" option of Netscape (see Figure 1.6). Other browsers also list all the sites you've visited during a session when you click a button next to the "Address" window. So even if you forget to write down an important address while you're at the site, its location should be preserved in these temporary records of your Internet travels.

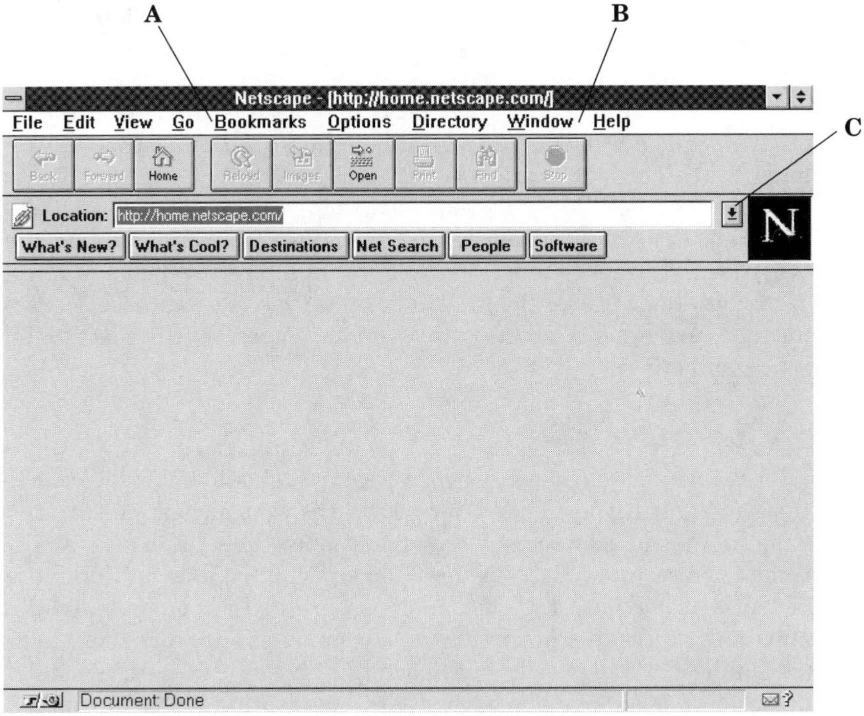

FIGURE 1.6 On Netscape and similar browsers, the "Bookmarks" (or "Hotlist") feature (A) will allow you to accumulate a personal list of important Internet addresses. Clicking on "Window" (B) and then "Window History" will reveal a list of all the sites you've visitied during a given session. The button next to the address window (C) will also open a box that records a short list of sites visited.

Source: Copyright 1996 Netscape Communications Corp. All Rights Reserved. This page may not be reprinted or copied without the express written permission of Netscape.

■ *The "Forward" and "Back" buttons.* You'll use these buttons more than any others. Like I said before, moving in the Web is a lot like driving in Boston: You keep leaving a rotary to explore some other part of town, and then return to the rotary to try another street leading to another part of town. As you move through the Web, you'll often click on hypertext links that lead to other links, but then you'll want to retrace your steps to get back to the original site. These two buttons will get you back and forth.

■ *The "Title" and "Address" window.* Key information resides here that you'll need not only to find the location of some information again but to cite it in your essay, if you use the information there. A

Web address will have something called a *URL* (uniform resource locator), which will look something like this:

http://www.unplug.com/great

or

gopher://tinman.mes.umn.edu:4242/11/other

Since URLs can look a little mind boggling, it's useful to be able to decode some key information from them right off.

—The first term—*http* or *ftp* or *gopher*—signals what part of cyberspace you've entered and what protocol you've used to get there. *Http* (or *hypertext transfer protocol*), for example, is Web space. A protocol is always followed by a colon and double-slashes that (unlike those in word-processing file names) lean to the right.

—The next series of characters represents the host computer—where the information is located. The key thing to notice is the *domains* at the ends of the addresses. The most common domains are *edu* (an educational institution), *com* (a commercial one), *org* (a nonprofit organization), *gov* (government), and *net* (a network administrative site). The domain characters provide vital information about the source of the information you've found.

—Finally, the last series of characters and slashes—which sometimes seem to go on endlessly—represent the *pathnames*. It's essential that when you write down a URL, you get *all* these characters, *exactly* as you seem them, including whether they are upper- or lowercase.

You'll learn a lot more about your browser the more you use it, but with this background knowledge, you're ready for a quick research foray on the Internet.

❑ *EXERCISE 1.5* A Quick Tour of the Internet

Launching a Subject Search

Step 1: Start Your Browser

If you're at a campus terminal, call up the browser on your school's network—probably Netscape or Mosaic. If you're at home, you may be able to call into the campus network using a modem in your computer; if not, you will likely have access to a commercial provider like Prodigy or America Online, each of which offers its own Web browser. Get your browser up and running. Once you do, you probably will be looking at the *home page* for your college or online service.

We'll begin with a subject search on your topic. This may seem an indirect way to find Internet sources for your essay, particularly if you can immediately think of some specific keywords you could use, like "method acting" or "Al-Anon" or "television violence." But a subject search can sometimes produce resources that a keyword search will not. Let's try it.

Step 2: Using the Virtual Library

A great place to begin a subject search on your topic is to visit the WWW Virtual Library. This is essentially a catalogue of Internet resources organized by subject. To get to it, type the following URL in the address window of your browser:

http://www.w3.org/pub/DataSources/bySubject/

What you will see next is a long list of subjects, alphabetically arranged and highlighted. Each of these provides links to other documents and locations on each subject. Look for the subject category that best matches your topic, and click on it with your mouse. For example, if my topic is method acting, I will notice that there is a "theater and drama" category. Try several subjects if one doesn't work. Then follow the links that seem promising for your topic.

If you find a document, reference source, or site that seems useful, collect the following information about it:

Author (if any): ______________________________

Title: ______________________________

Publication name and date of print version (if any): ______________

____

Online publication name or database (e.g., Boston Globe Online, *Oxford Text Archives, Alzheimer's Disease Research Center, etc.):* ____

____

Publication information of online version (volume or issue, date, page or paragraph numbers, or if none, n. pag.*):* ______________

____

Date you accessed it (day, month, year): ______________

Internet address: ______________________________

Also make sure to add it to your "Hotlist" or "Bookmarks."

Other Subject Search Sites

Several other servers are useful for Web subject searches. Give these a try:

http://www.einet.net/galaxy.html

Galaxy has a fairly elaborate subject hierarchy and, unlike the Virtual Library, provides a way for users to continually add new sites.

http://www.yahoo.com/

Yahoo is one of the most popular search engines, offering both subject trees to follow and keyword searching.

http://www.clearinghouse.net/

The Argus Clearinghouse bills itself as a premier Internet resource for subject-oriented Internet resource guides. Librarians are involved in this server.

Launching a Keyword Search

One of the great things about a subject search is that it often provides a context for the interesting documents you find; you can see a bunch of related sources, organizations, and sites. But sometimes, you know exactly what you're looking for: the name of an organization, the title of a document, or a series of words that nicely describes your topic. For example, Dave was writing a research essay on motorcycle helmet laws, a topic that didn't neatly fit under any one subject category. Doing an Internet search using keywords like "motorcycle helmet laws," or any number of variations on this phrase, would seem more promising than doing a subject search.

A growing number of Web search engines specialize in keyword searches. We'll preview just a few of the most popular and comprehensive here. Keep in mind, though, that the success of your Internet search, like your search in the university library, is strongly related to how well you choose search terms. In the library, you can consult the *Library of Congress Subject Headings* volumes to give you some guidance. However, the Internet provides no such standard for search terms. The key, then, is to have a range of alternative keywords to use in your search and to understand exactly how a particular search engine allows you to narrow your hunt.

Query, Query, Quite Contrary

While search engines may differ a bit in how they interpret keyword requests (something you should check on by clicking on the "Help" button before you launch a search), most use Boolean operators: AND, OR, NOT. (These are the same operators you'll use when searching a CD-ROM index at the library.) In addition, a number of search engines have another feature: the ability to search using a phrase relevant to a topic.

Knowing your Boolean can make life easier. It will make your Web searching quicker and the hits (matches) a search generates more useful. Let's use Dave's motorcycle helmet law project to illustrate.

If I type *motorcycle helmet laws* in the search field of WebCrawler, a popular search engine, it will look for Web sources that include *motorcycle, helmet,* or *laws* in their titles, URLs, and document contents. (The same thing, by the way, can be accomplished by typing an OR between each term. In this case, omitting an operator between terms *implies* the OR.) This query will produce a long list of sources—I turned up 16,604—and most of them will not be what Dave wants. Suppose I retype his request this way: *motorcycle and helmet and laws.* (You need not put the AND in caps.) Now, most search engines will only look for sources that include *all* of those terms, though not necessarily together in the same phrase. The list will be shorter—27 hits—and more useful. Another alternative is to use quotation marks around the request—"motorcycle helmet laws"—which, in most cases, tells the search engine to retrieve only sources that include all three terms *in the quoted phrase* where they appear together. Such a *literal search* can be an enormously productive method for finding useful information quickly if your topic, like Dave's, is often described in a phrase.

Unfortunately, WebCrawler turned up only 3 hits on Dave's topic with a literal search, and none seemed that promising. The second approach—using AND to connect the terms—turned out the best so far. Boolean operators can also be used together. For example, Dave could launch a search using these keywords: *motorcycle and helmet and laws or legislation.*

While AND, OR, and NOT are the usual Boolean operators, a number of popular search engines are now using the + sign, placed directly in front of a term. Like the Boolean AND, + signals that the term *must* be present in retrieved documents. The – sign, also placed before a word, would function as a NOT operator, indicating a term that must *not* be present in a document (e.g., *+Olympics –Atlanta*). Parentheses can also frequently be substituted for quotation marks in a literal search. Not using any punctuation between terms usually indicates the same thing as OR in this scheme—that any of the terms can be present. Using this system, then, Dave's search requests could be rewritten from *motorcycle and helmet and laws* to *+motorcycle +helmet +laws* and from *"motorcycle helmet laws"* to *(motorcycle helmet laws).*

Sometimes a search engine like MetaCrawler simplifies things by just asking whether you want to search terms as a phrase, as all of the words, or as any of the words. You just click on the answer without messing with Boolean operators.

There is much more to know about Boolean searches, but our discussion will end here, after covering the basics. Check your search engine "Help" files to learn more advanced techniques.

Step 3: Compose Keyword Searches

Using what you've learned so far about how to compose a keyword search, write three different possible arrangements of terms or phrases that seem promising for conducting a Web search on your topic. Consider promising synonyms or unusual terms that might help narrow your search. And make sure you're spelling things correctly.

1. ______________________________

2. ______________________________

3. ______________________________

You can fiddle with these keywords later, after you've tried them out. You'll likely get more ideas for searchable terms from the documents you scrounge up.

Step 4: Search on MetaCrawler

At last count, there were more than fifty search engines for roaming the Web. Which one should you start with? Why not start with a site that will launch a search on nine of the most popular search tools simultaneously? That's what MetaCrawler does. A single query on MetaCrawler will search the databases of Open Text, Lycos, WebCrawler, InfoSeek, Excite, Inktomi, Alta Vista, Yahoo, and Galaxy. It will then return what it deems the twenty (or more, if you want) most relevant hits from those databases and even eliminate duplicate listings. MetaCrawler and other metasearchers like it are the closest Web searching comes to one-stop shopping.

Use your browser to go MetaCrawler's search page, which is located at the following address:

http://www.metacrawler.com

Now, launch keyword searches on your topic. Note that you may have to rewrite your terms a bit from how you listed them in Step 3, since MetaCrawler does not use the Boolean operators AND, OR, and NOT but + (instead of AND) and – (rather than NOT). OR is implied when no syntax is used between terms. Parentheses around a phrase also substitute for quotation marks. Click on MetaCrawler's "Help" feature for more details.

Maria, a nursing major, worked with elderly patients at a local facility. Her research topic was Alzheimer's disease. She wisely chose to narrow her MetaCrawler search to find documents on diagnosis of the illness. Her search query, then, looked like this: *+diagnosis +Alzheimer's.* Of the twenty references returned, most were relevant, making this a highly successful search.

If MetaCrawler generated relevant hits on your topic, go to the most promising document and collect as much of the following information about it as you can:

Author (if any): ______________________________

Title: ______________________________

Publication name and date of print version (if any): ______________

____

Name of online publication or database: ______________

____

Online publication information (volume or issue, date, page or paragraph numbers, or if none, n. pag.*):* ______________

____

Date you accessed it (day, month, year): ______________

Internet address: ______________________________

MetaCrawler isn't the only so-called metasearch or parallel search engine, though it's arguably the best. Savvy Search (http://www.cs.colostate.edu/~dreiling/smartform.html) may be a useful alternative to MetaCrawler because it also searches FTP sites and newsgroups.

Step 5: Choosing a Single Search Engine

While MetaCrawler and other parallel searchers would seem to eliminate the need to use any one search tool, it's always wise to hedge your bets on the Internet. Never trust that any single search site—even a metasearch site—will give you complete coverage. There's also another reason to try one of the more conventional search engines: MetaCrawler may not exploit all the query features available on a particular engine.

Before we end this quick tour, try using one or more of the search engines (even it's one of MetaCrawler's nine) available on the Web to find information on your topic. (See the following box, "Choos-

CHOOSING A SEARCH ENGINE

A growing number of servers claim to have tamed the Wild West of cyberspace, each promising to help Internet researchers find relevant information faster and with less hassle. These so-called search engines, at least at first glance, seem to be remarkably similar. So how do you choose which one to use? Compare them using the following criteria:

- *The size of the database.* Obviously, you want to try out the search engine that will look at the largest number of documents and sites. Currently, Alta Vista or Excite, depending on whose claims you believe, is tops here. Alta Vista's collection includes over 21 million documents. Excite claims it can search 50 million Web sites and 10,000 newsgroups. Lycos also has a large database—over 5 million. Infoseek might also be worth trying; while it's database isn't quite as large, its coverage includes some unusual sources like newspapers, Medline, and commercial databases.
- *The quality of the indexing.* How does a search engine examine documents for a match with your query? Most examine document titles and URLs, but according to *PC World* magazine, several engines can do much more. Open Text indexes every word on every page of a document. Excite uses something called a *concept search,* which may return relevant documents even if there isn't a keyword match. Find out exactly how a particular search engine indexes Web material by consulting the "Help" file.
- *Query features.* You want a search engine that allows you to use a range of methods for altering or narrowing your query. Again, Alta Vista is strong on this quality and so is Open Text. Yahoo, one of the most popular servers, has a number of query features, including a subject catalog; it is also one of the easiest engines to use.

ing a Search Engine," for ways to evaluate search tools.) I recommend Alta Vista or Yahoo, but you may have your own preference. Choose one and write its name down here:

Search engine: ________________________________

You can find search tools like Yahoo on your browser by clicking on the appropriate button (for Netscape, it's "Net Search" on the toolbar), or you can go to all-in-one pages that feature links to most search engines if not forms for launching searches on any number of engines from the page itself. Of the many of these that exist, here are a few of the all-in-one search pages I've found useful:

http://www.albany.net/allinone/all1www.html

The "All-In-One Search Page" features interfaces with fifty Web search engines, including every one I've mentioned so far.

http://www.utexas.edu/search/

This site is sponsored by the University of Texas and organizes search engines in helpful categories, like those most appropriate for subject searches and keyword searches, indexes and abstracts, and libraries and universities. This page also includes special tools for searching non-Web sites, like FTP, WAIS, Usenet, and Telnet.

http://www.search.com/

This is another page loaded with the most popular search engines as well as a tool for deciding which one is appropriate for your topic.

http://www.nova.edu/Inter-Links/web/wwwsearch.html

Inter-Links is a great site that offers not only nine useful Web search engines but also tools for searching other space, like Telnet or Gopher. It also includes links to useful reference sources and libraries.

Launch a keyword search on your topic using the engine you have selected, and jot down information on a promising document below. Remember to keep trying different queries if one doesn't produce useful sources.

Author (if any): ______________________________

Title: ______________________________

Publication name and date of print version (if any): ______________

____

Name of online publication or database (if any): ______________

____

Online publication information (volume or issue, date, page or paragraph numbers, or if none, n. pag.*):* ____________________

____

Date of access: ______________________________

Internet address: ____________________________

Now you've got a taste of the Web's possibilities. If you're already hooked, beware: Don't give up on more conventional sources of information, especially library research. At the moment, the information available on the Internet is simply insufficient or sufficiently unreliable to be an exclusive source for your essay. If you're frustrated, don't give up on the Net. It may yet transform how we find information.

Considering Other Nonlibrary Sources: Interviews and Surveys

Before you prepare to begin library or Internet research in earnest next week, consider two alternative sources of information that may be extremely valuable: interviews and informal surveys. When I was writing my book on lobsters, I did lots of library research, particularly on the biology of the animal. I read studies on how female lobsters choose mates (they like those with the most appealing urine smell), examined charts on how fast lobsters grow and how old they are when they end up on my plate (about seven years), and consulted research by fisheries experts on rates of exploitation by the lobster industry.

I was especially interested in that last topic because one of the questions that triggered my search was wondering whether lobsters were being overfished. I learned about recruitment failure and fishing effort and the relationship between carapace length and maturity. Based on what I read, I concluded that almost all legal-sized lobsters are caught by fishermen every year and that at that size, they haven't had a chance to reproduce. But it wasn't until I stopped reading and interviewed a man named Guy Marchessault, a government biologist, that the meaning of all this became clear. "It's almost as if you drafted all thirteen-year-old, prepubescent kids and shot them all," he said. "How will you maintain a population of people if you don't let them reproduce? That's the conundrum with respect to lobsters."

Not only did talking to Guy clear up my own confusion, but he provided me with a memorable quote that would do the same for my readers, too.

You can find people who are experts in almost every conceivable field right on your own campus. And you'll probably find more experts who can be reached by phone with just a little digging. (See "Finding Experts" in Chapter 2 for information on arranging interviews.) The Internet also creates new opportunities for interviewing experts by e-mail. You may also know people who may not be authorities in the field you're researching but who have relevant experience to share. For example, André was researching a paper on banning fraternities and interviewed a fraternity president whose house was threatened with closure. Another student, Taylor, explored the impact of divorce on father-daughter relationships by interviewing her father.

When interviewing is expanded to include a group of people, usually by asking each person a standard set of questions, you are conducting an informal survey. Your results will not be statistically valid, but you may get a sense of what people think about some aspect of your research topic. The range of responses you receive may be illuminating, as well.

Very few topics don't lend themselves to interviews or surveys. However, you must determine whom to talk to about your topic. Who might be an expert and have useful information, or who has experience that is relevant? Jot down some tentative ideas in your research journal, and discuss them with your instructor next week.

2

□ ■ □

The Second Week

NARROWING THE SUBJECT

It never occurred to me that photography and writing had anything in common until I found myself wandering around a lonely beach one March afternoon with a camera around my neck. I had a fresh roll of film, and full of ambition, I set out to take beautiful pictures. Three hours later, I had taken only three shots, and I was definitely not having fun. Before quitting in disgust, I spent twenty minutes trying to take a single picture of a lighthouse. I stood there, feet planted in the sand, repeatedly bringing the camera to my face, but each time I looked through the viewfinder, I saw a picture I was sure I'd seen before, immortalized on a postcard in the gift shop down the road. Suddenly, photography lost its appeal.

A few months later, a student sat in my office, complaining that he didn't have anything to write about. "I thought about writing an essay on what it was like going home for the first time last weekend," he said, "but I thought that everyone probably writes about that in Freshman English." I looked at him and thought about lighthouse pictures.

Circling the Lighthouse

Most every subject you will choose to write about for this class and for this research paper has been written about before. The challenge is not to find a unique topic (save that for your doctoral dissertation) but to find an angle on a familiar topic that helps readers to see what they probably haven't noticed before. In "Why God Created

Flies," Richard Conniff took the most common of subjects—the housefly—and made it seem new by giving us a close look at its unusual habits and the surprising things he had to say about how the fly may be punishment for our own arrogance.

I now know that I was mistaken to give up on the lighthouse. The problem with my lighthouse picture, as well as with my student's proposed essay on going home, was not the subject. It was that neither of us had yet found our own angle. I needed to keep looking, walking around the lighthouse, taking lots of shots until I found one that surprised me, that helped me see the lighthouse in a new way, in *my* way. Instead, I stayed put, stuck on the long shot and the belief that I couldn't do better than a postcard photograph.

It is generally true that when we first look at something, we mostly see its obvious features. That became apparent when I asked my Freshman English class one year to go out and take pictures of anything they wanted. Several students came back with single photographs of Thompson Hall, a beautiful brick building on campus. Coincidentally, all were taken from the same angle and distance—straight on and across the street—which is the same shot that appears in the college recruiting catalog. For the next assignment, I asked my students to take multiple shots of a single subject, varying angle and distance. Several students went back to Thompson Hall and discovered a building they'd never seen before, though they walk by it every day. Students took abstract shots of the pattern of brickwork, unsettling shots of the clock tower looming above, and arresting shots of wrought iron fire escapes, clinging in a tangle to the wall.

The closer students got to their subjects, the more they began to see what they had never noticed before. The same is true in writing. As you move in for a closer look at some aspect of a larger subject, you will begin to uncover information that you—and ultimately your readers—are likely to find less familiar and more interesting. One writing term for this is *focusing*. (The photographic equivalent would be *distance from the subject.*)

From Landscape Shots to Close-Ups

When you wrote research reports in high school, you were a landscape photographer, trying to cram into one picture as much information as you could. A research report is a long shot. The college research essay is much more of a close-up, which means narrowing the boundaries of your topic as much as you can, always working for a more detailed look at some smaller part of the landscape.

Of course, you are not a photographer, and finding a narrow focus and fresh angle on your research topic is not nearly as simple as it might be if this were a photography exercise. But the idea is the same. You need to see your topic in as many ways as you can, hunting for the angle that most interests you; then go in for a closer look. One way to find your *focus* is to find your *questions.*

❑ *EXERCISE 2.1* Finding the Questions

Though you can do this exercise on your own, your instructor will likely ask that it be done in class this week. That way, students can all help each other. (If you do try this on your own, only do Steps 3 and 4 in your research notebook.)

Step 1: Take a piece of paper or a large piece of newsprint, and post it on the wall. At the very top of the paper, write the title of your tentative topic (e.g., *Legalization of Marijuana*).

Step 2: Take a few minutes and briefly describe why you chose the topic.

Step 3: Spend five minutes or so and briefly list what you know about your topic already (e.g., any surprising facts or statistics, the extent of the problem, important people or institutions involved, key schools of thought, common misconceptions, observations you've made, important trends, major controversies, etc.).

Step 4: Now spend fifteen or twenty minutes brainstorming a list of questions *about your topic* that you'd like to answer through your research. Make this list as long as you can; try to see your topic in as many ways as possible. Push yourself on this; it's the most important step.

Step 5: As you look around the room, you'll see a gallery of topics and questions on the walls. At this point in the research process, almost everyone will be struggling to find her focus. You can help each other. Move around the room, reviewing the topics and questions other students have generated. For each topic posted on the wall, do two things: Add a question *you* would like answered about that topic that's not on the list, and check the *one* question on the list you find most interesting. (It may or may not be the one you added.)

□ ■ □

If you do this exercise in class, when you return to your newsprint, note the question about your topic that garnered the most interest. This may not be the one that interests you the most, and you may choose to ignore it altogether. But it is helpful to get some idea of what typical readers might most want to know about your topic.

You also might be surprised by the rich variety of topics other students have tentatively chosen for their research projects. The last time I did this exercise, I had students propose papers on controversial issues like the use of dolphins in warfare, homelessness, the controversy over abolishment of fraternities, the legalization of marijuana, and the censorship of music. Other students proposed somewhat more personal issues, such as growing up with an alcoholic father, date rape, women in abusive relationships, and the effects of divorce on children. Still other students wanted to learn about more historical subjects, including the role of Emperor Hirohito in World War II, the student movement in the 1960s, and the Lizzie Borden murder case. A few students chose topics that were local. For example, one student whose great uncle was a union organizer wanted to investigate a strike that occurred seventy years ago at the old woolen mill in his hometown. Another student investigated a murder that took place on the Isle of Shoals, an island off the New Hampshire coast, a hundred years ago. Yet another explored his sons' struggle with attention-deficit disorder (ADD).

If the topic you've tentatively chosen is broad—such as abortion, whales, or child abuse—then Exercise 2.1 may help you discover the questions needed to narrow your topic into something more manageable. For example, if you're considering a paper on abortion, you'll quickly see how many angles there are on such a complicated subject: What is the history of abortion rights in the United States? How will the conservative majority on the Supreme Court influence *Roe* v. *Wade?* What is the impact of the abortion issue on local election campaigns? Each one of these questions could easily be answered in a ten-page paper, and most of these topics could likely be narrowed further.

Knowing the many questions your research project *could* answer is a start. But you obviously need to limit how many questions to ask, too. Once you've unleashed the many possibilities, you must harness a few in the service of your paper. If you need some help with that, try the next exercise.

❑ *EXERCISE 2.2*
Finding the Focusing Question

Review the questions you or the rest of the class generated in Exercise 2.1, Steps 4 and 5, and ask yourself, Which questions on the list am I most interested in that could be the focus of my paper? Remember, you're not committing yourself yet.

Step 1: Write the *one* question that you think would be the most interesting focus for your paper on the top of a fresh piece of newsprint or paper. This is your *focusing question.*

Step 2: Now build a new list of questions under the first one. What else do you need to know to answer your focusing question? For example, suppose your focusing question is, *Why do some colleges use unethical means to recruit athletes?* To explore that focus, you might need to find out:

Which colleges or universities have the worst records of unethical activities in recruiting?

In which sports do these recruiting practices occur most often? Why?

What are the NCAA rules about recruiting?

What is considered an *unethical practice?*

What efforts have been undertaken to curb bad practices?

Many of these questions may already appear on the lists you and the class generated, so keep them close at hand and mine them for ideas. Examine your tentative focusing question carefully for clues about what you might need to know. See also the box on pages 80–81, "Focusing: A Case Study," which describes how one student completed this exercise.

□ ■ □

Choosing a Trailhead

The importance of finding a narrow focus early can't be overestimated. I asked a reference librarian recently what she thought was the most common problem with student research papers. "That's easy," she said. "Students haven't narrowed their topics sufficiently. They come to us and ask where to begin looking for information on air pollution, and I could send them to any one of a hundred places."

Your research will be much more efficient if you have a limited focus, allowing you to concentrate on perhaps five relevant articles rather than wading through fifty about the broader topic. And by taking a closer look at a smaller part of your subject, you're much more likely to encounter information that will surprise you and your readers.

Settling on *one* central question that your research paper will attempt to answer is key. I call this a *trailhead question,* or the one that provides a path into your subject. It is just one of many paths, but it's the one that—at least for now—you're most interested in following.

FOCUSING: A CASE STUDY

Al, a student working on the topic children of alcoholics, came up with the following questions from Exercise 2.2:

Which family members are most affected by the drinker?

What can they realistically do to encourage the drinker to stop?

How likely is it that I will drink because my father did?

How effective is AA? Al-Anon? ACOA?

What's the relationship among these organizations?

What do they mean by "tough love?"

Do teenage children of alcoholics have any special problems?

How do the reactions of children differ?

Does a family share any of the responsibility for the alcoholic's disease?

Does it matter whether the father or mother is the drinker, or are the problems for the family the same in both cases?

What do they mean by "adult child?"

Al, the son of an alcoholic, had plenty of good questions. He had already done some reading on the subject, which helped, but he wasn't quite sure what he wanted to focus on. One question intrigued him the most—*Do teenage children of alcoholics have any special problems?*—but he was afraid the focus was too narrow. Would he be able to find enough information? Al was also interested in the last question, the meaning of the term *adult child.* But was that subject too broad? He was aware that scores of books had been published on adult children of alcoholics in recent years and thought he might be swamped by information.

If you can, start with the narrower focus. As you immerse yourself in research, you can broaden it a bit if you're having difficulty finding information, or you may encounter some interesting material that encourages you to change the focus altogether. If you start with a closer look at some aspect of your subject, your research efforts will be more efficient and the information you unearth will more likely surprise you.

Al tentatively chose the narrower focusing question: *Do teenage children of alcoholics have any special problems?* Working from that main question, he generated an additional list of questions that he may need to explore:

> Does anything about being a teenager make alcoholism harder?
>
> What do I remember from my own experience with my father at that age?
>
> Do groups like Al-Anon and ACOA have special meetings for teens?
>
> How do teenagers with alcoholic parents deal with the pressures to drink themselves?

Armed with this focus and some ideas about what to look for, Al was ready to begin his research in earnest.

As you follow it into the library this week, you may encounter other questions and other trails that take you off in new directions. That's fine. But at least have an initial direction.

What's Your Purpose?

Sometimes with high school research reports, it didn't seem to matter what you thought about the subject you were writing about. It seemed that what really mattered was whether you followed the proper format, cited sources correctly, and had a long bibliography. Though following the technical conventions is important in the college research paper, what matters even more is what you do with the information you find. Your paper must have a *purpose.*

Do You Have a Thesis?

In a broad sense, the purpose of your paper might be *to persuade, to analyze, to describe,* or *to explain* some aspect of your subject. Your purpose even might be a combination of all four. More specifically, your purpose in writing this paper is to find out *what you want to say* about your topic—in other words, finding your *thesis,* or your main idea—and then using persuasion, analysis, description, narration, or exposition to make that idea convincing to someone else. You may have learned that you need to define your thesis before you begin the research. That might be appropriate if you already know what you think. Last semester, I had a student, Kate, who believed strongly in the legalization of marijuana; she wanted to write a persuasive paper arguing that limited legalization was workable. Kate started with a thesis and went from there.

You might have a more tentative notion of what your paper's point will be, based on your current understanding of the topic. For example, the student investigating children of alcoholics, Al, believed that the answer to his focusing question—*Do teenage children of alcoholics have any special problems?*—might be that they do. That's what his own experience seemed to tell him. It's possible that the research will convince Al otherwise, or finding new information may help him make his thesis more specific. Perhaps the controlling idea of Al's paper might later become the assertion that teenage children of alcoholics have particular difficulty maintaining relationships later in life.

You may have chosen your topic precisely because *you don't know what you think*. So how do you know what you want to say until you've learned what others have said and had the chance to explore your topic? What I love about research is that it is a process of discovery. That demands an openness to what you'll find and a willingness to change your mind.

If, at this point, you can state your thesis, it would be useful to do so. But don't be inflexible. As with my students and their pictures of Thompson Hall, the more you look, the more you'll see. Be open to surprise.

If you can't state your thesis yet, don't worry. You can nail that down later, after you've done some digging. (See Exercise 4.1, "Reclaiming Your Topic," in Chapter 4.) But as you begin to mine your topic for information this week, always keep two questions in mind: What do I think? and What do I want to say?

❑ *EXERCISE 2.3*
Charting Your Course

You're now standing at the trailhead, ready to begin the journey into researching your topic. Before you do, in your research notebook, jot down the answers to the following questions. They'll help you get your bearings. Your instructor may ask you to bring this information to class or conference this week.

- What is my tentative focusing question?
- What other questions might help me explore that focus?
- What is my tentative thesis?

DEVELOPING A RESEARCH STRATEGY

Library Research Strategy

If you're finding questions about your topic that light small fires under you, then you may feel ready to plunge headlong into research this week. Before you do, plan your attack. At this point, you will likely return to the library with a stronger sense of what you want to know than you had last week, when you worked through the library exercise. But keep in mind that you still have many trails to follow and a formidable mass of information to consider. Where should you begin? How do you know where to go from there?

Moving from General to Specific

Often, researchers move from general sources to those that give topics more specific treatment. Look at it this way: You're not likely much of an authority on your topic yet, but by the time you write the final draft of your paper, you will be. What you have going for you at this point is your own curiosity. You're going to teach yourself about child abuse or the effects of caffeine, starting with the general knowledge you can find in the *New York Times* or *Redbook* and moving toward more specialized sources, such as the *Journal of Counseling Psychology* and the *New England Journal of Medicine*. Plan to begin with the general sources, as you did in Week 1 when you surveyed the landscape of your subject with the general encyclopedia and the *Reader's Guide*. Then move to progressively more specialized indexes and materials, digging more deeply as you become more and more of an expert in your own right.

Visualize books and articles—and the indexes that will help you find them—as occupying some spot on an inverted pyramid. Those sources toward the top are often written by nonexperts for a general audience. Those sources toward the bottom are more likely written by experts in the field for a more knowledgeable audience. An inverted pyramid of sources is shown in Figure 2.1.

Consider beginning with these generalized references, which will lead you to sources intended for a popular audience:

- General encyclopedias, almanacs, and dictionaries (e.g., *Encyclopaedia Britannica, Webster's Dictionary*)
- Indexes to popular periodicals and newspapers (e.g., *Reader's Guide,* InfoTrac, Newsbank)
- Card catalog (for popular books)

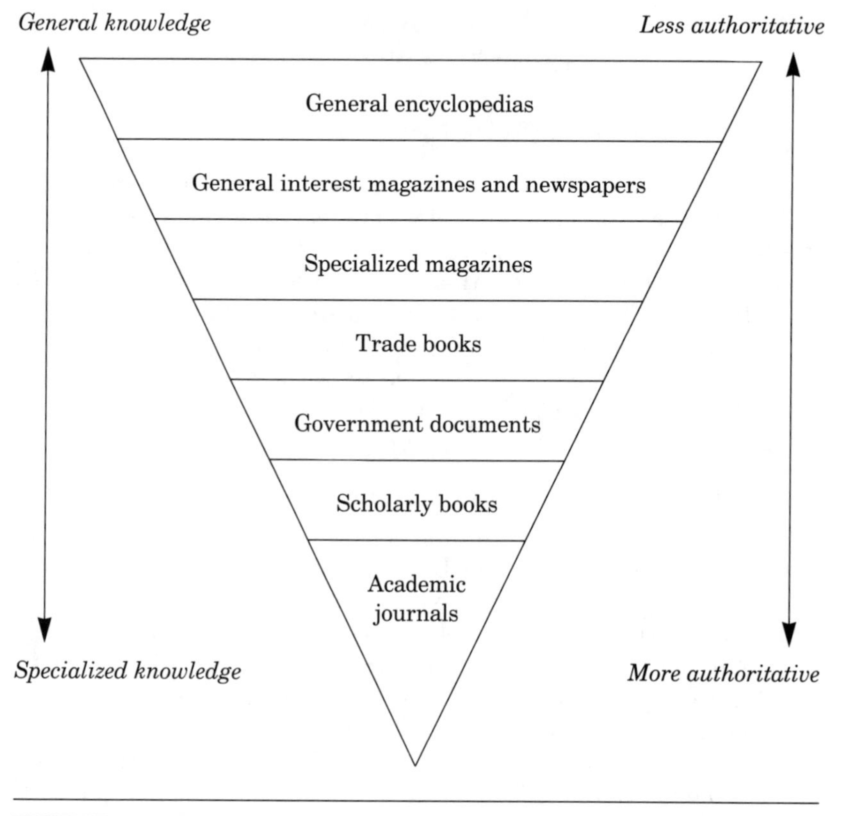

FIGURE 2.1 Pyramid of library sources

After consulting these sources, move to more specialized references, which will often lead you to books and articles written by experts on your topic for a more knowledgeable audience:

- Specialized encyclopedias and dictionaries (e.g., *Encyclopedia of Religion and Ethics, Dictionary of Philosophy and Psychology*—see "Second-Level Searching," Chapter 3)
- Specialized fact books (e.g., *Statistical Abstract of the United States, Facts on File*—see "Finding Quick Facts," Chapter 5)
- Academic indexes (e.g., the *Humanities* and *Social Science Indexes,* ERIC, Psyclit, and other CD databases)
- Card catalog (for more authoritative books on your topic)
- Government documents
- Biographical indexes
- Bibliographic indexes (see "Third-Level Searching," Chapter 3)
- Dissertation abstracts (see "Third-Level Searching," Chapter 3)

You may find as you work that this progression from the more general to the more specialized treatment of your topic occurs quite naturally. You'll quickly find that articles in popular magazines, newspapers, and books tell you what you already know. Your hunger to unearth new information will inevitably lead you to more specialized indexes, and you'll be able to read the articles they lead you to with more understanding as you become more of an authority on your topic.

Evaluating Library Sources

The aim of your research strategy is not only to find interesting information on your topic but to find it in *authoritative* sources. What are authoritative sources? In most cases, they are the most current sources. (The exception may be sources on historical subjects.) Authoritative sources are also those types found on the bottom of the pyramid (see Figure 2.1).

In part, the kinds of sources you rely on in preparing your paper depend on your topic. Sandra has chosen as her tentative focusing question, *What impact will the 1996 presidential election have on campaign finance reforms?* Because Sandra's topic addresses recent public policy, she'll likely find a wealth of information in newspapers and magazines but not much in books. She certainly should check the academic indexes on this topic—a CD database called *PAIS,* or Public Affairs Information System, would be a good bet—because it's likely that political scientists have something to say on the subject.

Why Journal Articles Are Better Than Magazine Articles. If your topic has been covered by academic journal articles, rely heavily on these sources. I've already mentioned that an article on, say, suicide among college students in a magazine like *Time* is less valuable than one in the *American Journal of Psychology.* Granted, the latter may be harder to read. But you're much more likely to learn something from a journal article because it's written by an expert and is usually narrowly focused. Also, because academic articles are carefully documented, you may be able to mine bibliographies for additional sources. And finally, scholarly work, such as that published in academic journals and books (usually published by university presses), is especially authoritative because it's subject to peer review. That means that every manuscript submitted for publication is reviewed by other authorities in the field, who scrutinize the author's evidence, methods, and arguments. Those articles that end up being published have truly passed muster.

Look for Often-Cited Authors. As you make your way through information on your topic, pay attention to names of authors whose work you often encounter or who are frequently mentioned in bibliographies. These individuals are often the best scholars in the field, and it will be useful to become familiar with their work and use it, if possible, in your paper. If an author's name keeps turning up, use it as another term for searching the card catalog and the indexes. Doing so might yield new sources you wouldn't necessarily encounter using subject headings. A specialized index—such as the *Social Science* and *Humanities Indexes*—will help you search for articles both by a particular author and by other authors who cite a particular author in their articles. (For more information on how to use the citation indexes, see "Using Citation Indexes" in Chapter 3.)

Primary over Secondary Sources. Another way of looking at information is to determine whether it's a *primary* or a *secondary* source. A primary source presents the original words of a writer—her speech, poem, eyewitness account, letter, interview, autobiography. A *secondary* source presents somebody else's work. Whenever possible, choose a primary source over a secondary one, since the primary source is likely to be more accurate and authoritative.

The subject you research will determine the kinds of primary sources you encounter. For example, if you're writing a paper on a novelist, then his novels, stories, letters, and interviews are primary sources. My topic on the engineering of the Chicago River in 1900, a partly historical subject, might lead me to a government report on the project or a firsthand account of its construction in a Chicago newspaper. Primary sources for a paper in the sciences might be findings from an experiment or observations and for a paper in business, marketing information or technical studies.

Not All Books Are Alike. When writing my high school research reports, I thought that a book was always the best source because, well, books are thick, and anyone who could write that much on any one subject probably knows what she's talking about. Naive, I know.

One of the things college teaches is *critical thinking*—the instinct to pause and consider before rushing to judgment. I've learned not to automatically believe in the validity of what an author is saying (as you shouldn't for this author), even if he did write a thick book about it.

If your topic lends itself to using primarily books as sources, then evaluate the authority of each before deciding to use it in your paper. This is especially important if your paper relies heavily on one particular book. Consider the following:

- Is the book written for a general audience or more knowledgeable readers?
- Is the author an acknowledged expert in the field?
- Is there a bibliography? Is the information carefully documented?
- How was the book received by critics? To find out, consider checking the following indexes, which may feature summaries of reviews or refer you to articles reviewing the book in magazines and journals. Entries are usually listed by author, title, or reviewer.

  Book Review Digest. New York: Wilson, 1905–date. *Provides summaries and quotations from reviews.*

  Current Book Review Citations. New York: Wilson. *Indexes reviews in over one thousand periodicals.*

Internet Research Strategy

In the last chapter—Exercise 1.4, "A Quick Tour of the Internet"—you got started with Internet research using the World Wide Web, a fairly accessible means of reaching all kinds of information in cyberspace. Because of its reach and ease of use, an Internet search strategy might begin with the Web. But keep in mind that Internet resources are much more vast than the WWW, and if you want to do a thorough Internet search (something that may or may not be useful, depending on your topic), you should learn to use some of the search tools appropriate for exploring each part of the cluttered canvas of cyberspace. That's especially true of the information-rich areas like Gopher space, FTP space, WAIS space, Telnet space, and newsgroups. (See Chapter 1, "A Cluttered Canvas," for an explanation of these Internet protocols.)

Many Web search engines can enter these spaces but don't look around thoroughly. A sophisticated Internet research strategy, then, might move from Web searching to more targeted searching. For example, *Veronica* and *Jughead* are useful tools for targeting searches of the thousands of files on a wide range of subjects available through Gopher. *Archie* is a useful tool for searching FTP space. Veronica, Jughead, and Archie are often available on campus servers.

Some of the gateways to these spaces are Web based. See Figure 3.5 in the next chapter (page 140), which lists several of these, including their Web addresses.

If the basic library research strategy is to move down the pyramid of sources, from the more general to the more specific, then the basic Internet research strategy is simply to cover as much ground as possible. Since no single indexing system or reference source of Net information is comprehensive, the essential strategy is to launch multiple searches, each of which looks carefully at a part of cyberspace that might reveal information about your topic.

Evaluating Online Sources

Evaluating online sources is tricky. The democratic nature of the Internet—almost anyone with a computer and modem can publish—is both its strength and its weakness. For the researcher, especially, this poses a basic problem: Can you trust the information you find on the Internet?

Maybe, maybe not. Here are some things to consider:

1. *Does the information originally come from an authoritative source?* Often, this is easy to determine. For example, the FTP archives of the federal government include files from most agencies and reports from their experts. The Oxford Text Archives include electronic versions of great works of literature, chosen by archive librarians and associated academics. The *New York Times Online* is the electronic version of what many people consider the most authoritative national newspaper. The *American Mathematical Society Bulletin,* a peer-reviewed print journal, also has an electronic version; the math articles you find there should be authoritative. Likewise, you should be able to trust the credibility of articles included in the growing number of peer-reviewed scholarly journals that only appear online, like the *Electronic Green Journal,* an environmental journal, and the *Murdoch Electronic Journal of Law.* The Association of Research Libraries offers an online *Directory of Electronic Journals and Newsletters,* which cannot only help you find a publication on your topic but can tell you its academic affiliation and whether it's peer reviewed. Check it out at gopher://arl.cni.org:70/scomm/edir/edir95.

2. *Is the source institutional?* If you're researching Alzheimer's disease and find your way to the Web page for the Center for Alzheimer's and Neurogenerative Disorders, then you can be reasonably certain that the information provided there is reliable, especially if you are able to find some basic information on the institution offering it. A description listing an organization's mission and affiliations is frequently available at its Web site; use that information to help you evaluate the organization's credibility as a source.

3. *If the document is from an individual, can you establish her expertise on your topic?* Frequently, you will encounter documents that have single authors with no clear affiliations or other credentials that you can use to establish their expertise. Authorless documents should automatically arouse your suspicion, unless you know they are origi-

nally from authoritative sources. Documents archived by educational institutions (the domain *edu* will appear in the Internet address) or authored by people who appear to be associated with universities *may* be more reliable, but it would be helpful to establish whether anyone else acknowledges these individuals' expertise to talk about your topic. There are a few ways to do this:

- Check and see if their names appear in the text or citations of any of the print sources you've collected on your topic.
- Run a search using their names in the library to see whether they've published books or articles on the subject.
- E-mail the authors and ask what else they've published. Many Web sites include the e-mail addresses of their authors, and you can send messages easily by clicking on the hypertext link. By initiating this contact, you might even open up the possibility of doing an e-mail interview. (See "The E-Mail Interview," later in this chapter.)

A good researcher always takes a skeptical view of claims made in print; she should be even more wary of claims made in Internet documents. And while these approaches for evaluating online sources should help, it still can be pretty tricky deciding whom to take seriously in cyberspace. So to sort it all out, always ask yourself these questions: How important is this Internet document to my research? Do I really need it? Might there be a more reliable print version?

Arranging Interviews

A few years ago, I researched a local turn-of-the-century writer named Sarah Orne Jewett for a magazine article. I dutifully read much of her work, studied critical articles and books on her writing, and visited her childhood home, which is open to the public in South Berwick, Maine. My research was going fairly well, but when I sat down to begin writing the draft, the material seemed flat and lifeless. A few days later, the curator of the Jewett house mentioned that there was an eighty-eight-year-old local woman, Elizabeth Goodwin, who had known the writer when she was alive. "As far as I know, she's the last living person who knew Sarah Orne Jewett," the curator told me. "And she lives just down the street."

The next week, I spent three hours with Elizabeth Goodwin, who told me of coming for breakfast with the famous author and eating strawberry jam and muffins. Elizabeth told me that many years after Jewett's death, the house seemed haunted by her friendly presence. One time, when Elizabeth lived in the Jewett house as a curator, some unseen hands pulled her back as she teetered at the top of the steep stairs in the back of the house. She likes to believe it was the author's ghost.

This interview transformed the piece by bringing the subject to life—first, for me as the writer, and then later, for my readers. Ultimately, what makes almost any topic compelling is discovering why it matters to *people*—how it affects their lives. Doing interviews with people close to the subject, both experts and nonexperts, is often the best way to find that out.

Last week, I urged you to consider whether interviews could be a part of your research, perhaps an integral part. If you have doubts, reconsider, because material from interviews may transform your paper in the way it did mine.

If you'd like to do some interviews, now is the time to begin arranging them.

Finding Experts

You may be hesitant to consider finding authorities on your topic to talk to because, after all, you're just a lowly student who knows next to nothing. How could you possibly impose on that sociology professor who published the book on anti-Semitism you found in the library? If that's how you feel, keep this in mind: *Most people, no matter who they are, love the attention of an interviewer, no matter who she is, particularly if what's being discussed fascinates them both.* Time and again, I've found my own shyness creep up on me when I pick up the telephone to arrange an interview. But almost invariably, when I get there and start talking with my interview subject, the experience is great for us both.

How do you find experts to interview?

- *Check your sources.* As you begin to collect books, articles, and Internet documents, note their authors and affiliations. I get calls from time to time from writers who came across my book on lobsters in the course of their research and discovered that I was at Boise State University. Sometimes the caller will arrange a phone interview or, if he lives within driving distance, a personal interview.

- *Check the phone book* The familiar Yellow Pages can be a gold mine. Carin, who was writing a paper on solar energy, merely looked under that heading and found a local dealer who sold solar systems to homeowners. Mark, who was investigating the effects of sexual abuse on children, found a counselor who specialized in treating abuse victims.

- *Ask your friends and your instructors.* Your roommate's boyfriend's father may be a criminal attorney who has lots to say about the insanity defense for your paper on that topic. Your best friend may be taking a photography course with a professor who would be a great interview for your paper on the work of Edward Weston. One of your instructors may know other faculty working in your subject area who would do an interview.

- *Check the faculty directory.* Many universities publish an annual directory of faculty and their research interests. On my campus, it's called the *Directory of Research and Scholarly Activities*. From it, I know, for example, that two professors at my university have expertise in eating disorders, a popular topic with student researchers.

- *Check the* Encyclopedia of Associations. This is a wonderful reference book that lists organizations with interests ranging from promoting tofu to preventing acid rain. Each listing includes the name of the group, its address and phone number, a list of its publications, and a short description of its purpose. Sometimes, these organizations can direct you to experts in your area who are available for live interviews or to spokespeople who are happy to provide phone interviews.

- *Check the Internet.* You can find the e-mail addresses and phone numbers of many scholars and researchers on the Internet, including those affiliated with your own and nearby universities. Often, these experts are listed in online directories for their colleges or universities. Sometimes, you can find knowledgable people by subscribing to a listserv or Internet discussion group on your topic. Occasionally, an expert will have her own Web page, and her e-mail address will provide a hypertext link. (For more details, see "Finding People on the Internet," later in this chapter.)

Finding Nonexperts Affected by Your Topic

The distinction between *expert* and *nonexpert* is tricky. For example, someone who lived through twelve months of combat in Vietnam certainly has direct knowledge of the subject, though probably he hasn't published an article about the war in *Foreign Affairs*. Similarly, a friend who experienced an abusive relationship with her boyfriend or overcame a drug addiction is, at least in a sense, an authority on abuse or addiction. Both individuals would likely be invaluable interviews for papers on those topics. The voices and the stories of people who are affected by the topic you're writing about can do more than anything else to make the information come to life, even if they don't have Ph.D.s.

You may already know people you can interview about your topic. Last semester, Amanda researched how mother-daughter relationships change when a daughter goes to college. She had no problem finding other women anxious to talk about how they get along with their mothers. A few years ago, Dan researched steroid use by student athletes. He discreetly asked his friends if they knew anyone who had taken the drugs. It turned out that an acquaintance of Dan's had used the drugs regularly and was happy to talk about his experience.

If you don't know people to interview, try posting notices on campus kiosks or bulletin boards. For example, "I'm doing a research project and interested in talking to people who grew up in single-parent households. Please call 555-9000." Also poll other students in your class for ideas about people you might interview for your paper. Help each other out.

Making Contact

By the end of this week, you should have some people to contact for interviews. First, consider whether to ask for a personal, telephone, or e-mail interview or perhaps, as a last resort, to simply correspond by mail. The personal interview is almost always preferable; you cannot only listen but watch, observing your subject's gestures and the setting, both of which can be revealing. When I'm interviewing someone in her office or home, for example, one of the first things I may jot down are the titles of books on the bookshelf. Sometimes, details about gestures and settings can be worked into your paper. Most of all, the personal interview is preferable because it's more natural, more like a conversation.

Be prepared. You may have no choice in the type of interview. If your subject is off campus or out of state, your only options may be the telephone, e-mail, or regular mail.

When contacting a subject for an interview, first state your name and then briefly explain your research project. If you were referred to the subject by someone she may know, mention that. A comment like "I think you could be extremely helpful to me" or "I'm familiar with your work, and I'm anxious to talk to you about it" works well. That's called *flattery,* and as long as it isn't excessive or insincere, we're all vulnerable to it.

It is gracious to ask your prospective subject what time and place for an interview may be convenient for her. Nonetheless, be prepared to suggest some specific times and places to meet or talk. When thinking about when to propose the interview with an expert on your topic, consider arranging it *after* you've done some research. You will not only be more well informed, you will have a clearer sense of what you want to know and what questions to ask. (For information on preparing for and conducting an interview, see "Conducting Interviews" in Chapter 3.)

The E-Mail Interview

The Internet opens up new possibilities for interviews; increasingly, experts (as well as nonexperts interested in certain subjects) are accessible through electronic mail (e-mail) and newsgroups. While electronic communication doesn't quite approach the conversational quality of the conventional face-to-face interview, the spontaneous nature of e-mail exchanges can come pretty close. It's possible to send a message, get a response, respond to the response, and get a further response—all in a single day. And for shy interviewers and interviewees, an e-mail conversation is an attractive alternative.

Finding People on the Internet. Finding people on the Internet doesn't have to involve a needle and hay if you have some information on whom you're looking for. If you know an expert's name, his organizational affiliation, and his geographical location, several search tools may help you track down his e-mail address, if he has one. But perhaps the easiest way to use the Net to find someone to interview is through a Web document on your topic. For example, when researching this new edition of this book, I encountered an online version of the Alliance for Computers and Writing's proposals for MLA-style electronic citations, authored by Janice Walker. Walker's e-mail address was a hyperlink in that document, so had I wished to ask her some questions, all I would have had to do was click on her name. Authors of Web pages frequently provide their addresses as links, inviting comments about their texts and the like. Thus, it seems safe to assume that they are probably willing to entertain questions from researchers, too.

Plucking an e-mail address from a Web page is the easiest way to find an interview subject. But what if you just have someone's name and organizational affiliation? A number of search tools specialize in looking for people on the Internet. Here are a few:

Four11—http://www.four11.com

This server claims to be the largest White Pages directory to the Internet, with over 6.5 million listings. You can search by name, location, old e-mail address, or group connection. I've had mixed success using Four11 for people searches.

Netfind—http://www.nova.edu/Inter-Links/netfind.html

A people query on Netfind involves the name and keywords that point to a person's location (e.g., "Ballenger Boise English"). Netfind then returns a list of directories that seem likely sites to search for the address. Unless you have some fairly specific information about your subject and his location, sifting through these directories can be time consuming.

College E-Mail Addresses—
http://www.qucis.queensu.ca/FAQs/email/college.html

If your potential interview subject is at a university, this database might be the place to start. It includes not only many U.S. universities but a number of foreign ones, as well. The database lists both student and faculty addresses.

Making Contact by E-Mail. Once you find the e-mail address of someone who seems a likely interview subject, proceed courteously and cautiously. One of the Internet's haunting issues is its potential to violate privacy. Be especially careful if you've gone to great lengths in hunting down the e-mail address of someone involved with your research topic; she may not be keen on receiving unsolicited e-mail messages from strangers. It would be courteous to approach any potential interview subject with a short message that asks permission for an online interview. To do so, briefly describe your project and why you thought this individual might be a good source for you. As always, you will be much more likely to get an enthusiastic response from someone if you can demonstrate your knowledge of her work on or experience with your topic.

Let's assume your initial contact has been successful and your subject has agreed to answer your questions. Your follow-up message should ask a *limited* number of questions—say, four or five—that are thoughtful and, if possible, specific. Keep in mind that while the e-mail interview is conducted in writing rather than through talking, many of the methods for handling conventional interviews still apply (see "Conducting Interviews" in Chapter 3).

Finding People on Listservs and Newsgroups. Another way to find interview subjects on the Internet is through *listservs* and *Usenet newsgroups.* These are electronic discussion groups, both academic and nonacademic, that people subscribe to when they share an interest in something; topics run well into the thousands, from mountain biking to medieval music to David Barry. I belong to several listservs, including one called *Megabyte University (MBU),* which talks about computers and writing. Subscribers determine the topics of discussion by sending e-mail messages to the list, which often prompt responses from others.

Anyone can post a question to a discussion group if he subscribes to the list, and in most cases, that's easy to do. (See the box "How to Subscribe to an Internet Discussion Group.") Moreover, since you're querying people who already know something about your topic, you're likely to get good information and quotable comments. You will, that is, if you ask good questions that demonstrate you've already done some thinking about your topic.

HOW TO SUBSCRIBE TO AN INTERNET DISCUSSION GROUP

To get on a listserv, you'll obviously need to have your own e-mail address, to which posts will be sent. You'll also need a browser (or other program) that can both read and send e-mail messages.

Every e-mail discussion list has two addresses: (1) the list address to which you send messages to other subscribers and (2) the listserv address from which you send requests to subscribe and unsubscribe. The distinction between these two addresses is important.

- *Subscribing*—To subscribe to a list, send an e-mail message with the following line:

SUB listname yourname

For example, to subscribe to *Studies in Antiquities and Mormonism,* or *SAMU-L,* I'd send a message to LISTSERV@BINGVMB.CC.BINGHAMTON.EDU that would simply say:

SUB SAMU-L Bruce Ballenger

- *Unsubscribing*—To unsubscribe to a list, send an e-mail message to the listserv address with this line:

SIGNOFF listname

If I tired of Antiquities and Mormonism, I'd send the following message to the listserv address:

SIGNOFF SAMU-L

- *Posting*—If I wanted to send a message to all the members of a list, I'd send it to the list address *rather* than the listserv address. For example, to send a question to the Mormonism and antiquities list, I'd send an e-mail to:

SAMU-L@BINGVMB.CC.BINGHAMTON.EDU

Querying a discussion group is not something to do to collect general or background information. People subscribe to these lists because they're anxious to participate in a stimulating conversation with others knowledgeable about the topic. While they often welcome research questions from students and others, subscribers may have little patience for questions that imply the researcher doesn't know the field or know what he's after. For example, several members of an environmental listserv complained that the list was "flooded" with "ill-expressed" questions from students, asking obvious and general questions like "What is hazardous waste?" This example should not discourage you from posting research questions to a discussion group, but do so only when you're ready to pose questions that have grown out of your reading on the topic.

Deciding What to Ask. Another way to get some help with knowing what to ask—and what not to—is to spend some time following the discussion of list participants before you jump in yourself. You might find, for example, that it would be far better to interview one participant with interesting views rather than to post questions to the whole list.

But if you do want to query the whole listserv or newsgroup, avoid posting a question that may have already received substantial attention from participants. You can find out what's been covered by consulting the list's *FAQ*s *(frequently asked questions).* The issue you're interested in may be there, along with a range of responses from list participants, which will spare you the need to ask the question at all. FAQs are often available through the discussion group at an FTP site, but one Web site that archives FAQs from most of the Usenet groups is located at http://www.cis.ohio-state.edu/hyptertext/faq/usenet/FAQ-List.html. I found, for example, a wonderfully informative collection of FAQs on acquired immune deficiency syndrome (AIDS), archived by the newsgroup sci.med.aids. Its FAQs would be a great place to start a research project on this topic.

Finding a Group on Your Topic. With thousands of discussion groups on every imaginable topic, how do you find one on your topic? You can use a number of Web search engines to retrieve articles from Usenet groups, but you should try some searchable indexes to Usenet space, too. Two I've found quite helpful are as follow:

Directory of Scholarly and Professional E-Conferences
http://n2h2.com/KOVACs/

Inter-Links Search List of Discussion Groups
http://www.nova.edu/Inter-Links/cgi-bin/lists

These tools will allow you to search an index of newsgroups by subject and then return information about those groups that match your query, including their subscription addresses (see Figure 2.2, pp. 98–99). Both of the directories just listed are particularly useful in finding academically oriented discussion groups.

Another Usenet search tool called *Deja News* (http://www.dejanews.com) has some nifty features. It allows you to search by subject or keywords and then produces a list of individual articles or postings that match your query from a number of newsgroups, which includes a hypertext link to the author of each post; this strikes me as an ideal way to find interview subjects through the Internet. Once you have chosen a post to read, *Deja News* will also generate links to all the other articles in that topic thread. A final feature that makes this Web site appealing is that it makes it possible for any user to post a message to any discussion group through the *Deja News* service. This feature can be particularly helpful if your browser doesn't have the software to read Usenet articles or post to them. And even if your browser has this capability, you'll find the *Deja News* posting feature convenient.

Planning Informal Surveys

Christine was interested in dream interpretation, especially exploring the significance of symbols or images that recur in many people's dreams. She could have simply examined her own dreams, but she thought it might be more interesting to survey a group of fellow students, asking how often they dream and what they remember about it. An informal survey, in which she would ask each person several standard questions, seemed worth trying.

You might consider it, too, if the responses of a group of people to some aspect of your topic would reveal a pattern of behavior, attitudes, or experiences worth analyzing. Informal surveys are decidedly unscientific. You probably won't get a large enough sample size, nor do you likely have the skills to design a poll that would produce statistically reliable results. But you probably won't actually base your paper on the survey results, anyway. Rather, you'll present specific, concrete information that *suggests* some patterns in your survey group, or at the very least, some of your own findings will help support your assertions.

Defining Goals and Audience

Begin planning your informal survey by defining what you want to know and whom you want to know it from. Christine suspected that many students have dreams related to stress. She wondered if there were any similarities among students' dreams. She was also

WWW view of Directory of Scholarly and Professional E Conferences

- What is Directory of Scholarly and Professional E-Conferences

You may explore the materials of this resource by any of the following:

- Subject-based listing
- Alphabetical listing

You may search the materials of this resource by any of the following:

Fields

Directory: [ ]
Discussion Name: [ ]
Topic: [Service learning]
Keywords: [ ]

- Number of matches permitted (default is 40): [40]
- Type of boolean operator used between fields: [and]

[search]

[reset form]

PRODIGY ® Web Browser: Directory of Scholarly and Professional E-Conferences (http://n2h2.com/KOVACS/)

FIGURE 2.2 The *Directory of Scholarly and Professional E-Conferences* allows you to search for discussion groups on your topic by subject or keyword. A search on "Service learning" produced a discussion group on the topic, including a subscription address. Especially useful is the address for the group's archives, which may be a rich source of information.
Source: Reprinted with permission of Kovacs Consulting.

Other Internet gateways to Directory of Scholarly and Professional E Conferences

GOPHER

- gopher://gopher.usask.ca

Directory:
Social Activism
Discussion Name:
SERVICE-LEARNING
Topic:
Discussion that is hosted by Communications for a Sustainable Future (CSF) at the University of Colorado at Boulder (CU), and by PEACE, the peace studies list and data-base at CSF. SERVICE-LEARNING is part of larger project that integrates the discussion group with a Service-Learning database - a "virtual" service-learning library - called SERVICE-LEARNING.
Subscription Address:
listproc@csf.colorado.edu
Moderated?
No
Archives:
ftp://csf.colorado.edu - gopher://csf.colorado.edu
Contact Address:
Robin Crews crews@csf.colorado.edu
Submission Address:
SERVICE-LEARNING@csf.colorado.edu
Keywords:
Service Learning
VR:
10th Revision 1/1/96

This page was generated by SFgate 4.0.29.

PRODIGY ® Web Browser: Wais document (plain text) (http://n2h2.com/cgi-bin/S... 2FKOVACS/HTML/929/1=)

FIGURE 2.2 Continued

curious about how many people remember their dreams, and how often, and if that might be related to gender. Finally, Christine wanted to find out whether people have recurring dreams and if so, what those were about. There were other things she wanted to know, too. But she knew she had to keep the survey short, probably no more than seven questions.

If you're considering a survey, make a list in your research notebook of things you might want to find out and specify what group of people you plan to talk to. College students? Female college students? Attorneys? Guidance counselors? Be as specific as you can about your target group.

Types of Questions

Next, consider what approach you will take. Will you ask *open-ended questions,* which give respondents plenty of room to invent their own answers? For example, Christine might ask, *Have you ever had any dreams that seem related to stress?* The payoff for an open-ended question is that sometimes you get surprising answers. The danger, which seems real with Christine's question, is that you'll get no answer at all. A more *directed question* might be, *Have you ever dreamed that you showed up for class and didn't know that there was a major exam that day?* Christine will get an answer to this question—yes or no—but it doesn't promise much information. A third possibility is the *multiple-choice question.* It ensures an answer and is likely to produce useful information. For example:

```
Have you ever had any dreams similar to these?
  a. You showed up for a class and didn't know
     there was a major exam.
  b. You registered for a class but forgot to
     attend.
  c. You're late for a class or an exam but can't
     seem to move fast enough to get there on
     time.
  d. You were to give a presentation but forgot
     all about it.
  e. None of the above.*
```

*Reprinted with permission of Christine Bergquist.

Ultimately, Christine decided to combine the open-ended question about stress and the multiple-choice approach, hoping that if one didn't produce interesting information, the other would (see Figure 2.3). She also wisely decided to avoid asking more than seven questions, allowing her subjects to respond to her survey in minutes.

Survey Design

A survey shouldn't be too long (probably no more than six or seven questions), it shouldn't be biased (asking questions that will skew the answers), it should be easy to score and tabulate results (especially if you hope to survey a relatively large number of people), it should ask clear questions, and it should give clear instructions for how to answer.

As a rule, informal surveys should begin as polls often do: by getting vital information about the respondent. Christine's survey began with questions about the gender, age, and major of each respondent (see Figure 2.3). Depending on the purpose of your survey, you might also want to know things such as whether respondents are registered to vote, whether they have political affiliations, what year of school they're in, or any number of other factors. Ask for information that provides different ways of breaking down your target group.

Avoid Loaded Questions. Question design is tricky business. An obviously biased question—*Do you think it's morally wrong to kill unborn babies through abortion?*—is easy to alter by removing the charged and presumptuous language. (It is unlikely that all respondents believe that abortion is killing.) One revision might be, *Do you support or oppose providing women the option to abort a pregnancy during the first twenty weeks?* This is a direct and specific question, neutrally stated, that calls for a yes or no answer. The question would be better if it were even more specific.

Controversial topics, like abortion, are most vulnerable to biased survey questions. If your topic is controversial, take great care to eliminate bias by avoiding charged language, especially if you have strong feelings yourself.

Avoid Vague Questions. Another trap is vague questions: *Do you support or oppose the university's alcohol policy?* In this case, don't assume that respondents know what the policy is unless you explain it. Since the campus alcohol policy has many elements, this question might be redesigned to ask about one of them: *The university recently established a policy that states that underage students caught drinking in campus dormitories are subject to eviction. Do you support or oppose this policy?* Other equally specific questions might ask about other parts of the policy.

The following survey contains questions about dreaming and dream content. The findings gathered from this survey will be incorporated into a research paper on the function of dreaming and what, if anything, we can learn from it. I'd appreciate your honest answers to the questions. Thank you for your time!

General Subject Information

Gender: ☐ Male ☐ Female

Age: ____________

Major: ____

Survey Questions

(circle all letters that apply)

1. How often do you remember your dreams?
   A. Almost every night
   B. About once a week
   C. Every few weeks
   D. Practically never
2. Have you ever dreamt that you were:
   A. Falling?
   B. Flying?
3. Have you ever dreamt of:
   A. Your death?
   B. The death of someone close to you?
4. Have you ever had a recurring dream?
   A. Yes
   B. No

   If yes, How often? ____________________
   What period of your life? ____________________
   Do you still have it? ____________________
5. Have you ever had any dreams similar to these?
   A. You showed up for a class and didn't know there was a major exam.
   B. You're late for a class or an exam but can't seem to move fast enough to get there.
   C. You were to give a presentation but forgot all about it.
6. Do you feel your dreams:
   A. Hold some deep, hidden meanings about yourself or your life?
   B. Are meaningless?
7. Please briefly describe the dream you best remember or one that sticks out in your mind. (Use the back of this survey.)

FIGURE 2.3 Sample informal survey

Source: Reprinted with permission of Christine Bergquist.

Drawbacks of Open-Ended Questions. Open-ended questions often produce fascinating answers, but they can be difficult to tabulate. Christine's survey asked, *Please briefly describe the one dream you best remember or one that sticks out in your mind.* She got a wide range of answers—or sometimes no answer at all—but it was hard to quantify the results. Most everyone had different dreams, which made it difficult to discern much of a pattern. She was still able to use some of the material as anecdotes in her paper, so it turned out to be a question worth asking.

Designing Multiple-Choice Questions. The multiple-choice question is an alternative to the open-ended question, leaving room for a number of *limited* responses, which are easier to quantify. Christine's survey had a number of multiple-choice questions.

The challenge in designing multiple-choice questions is to provide choices that will likely produce results. From her reading and talking to friends, Christine came up with what she thought were three stress-related dreams college students often experience (see question 5, Figure 2.3). The results were interesting (45 percent circled "B") but unreliable, since respondents did not have a "None of the above" option. How many respondents felt forced to choose one of the dreams listed because there was no other choice? Design choices you think your audience will respond to, but give them room to say your choices weren't theirs.

Continuum Questions. Question 6 (see Figure 2.3) has a similar choice problem in that it asks a direct either/or question: *Do you feel your dreams: (A) Hold some deep, hidden meanings about yourself or your life? or (B) Are meaningless?* Phrased this way, the question forces the respondent into one of two extreme positions. People are more likely to place themselves somewhere in between.

A variation on the multiple-choice question is the *continuum,* where respondents indicate how they feel by marking the appropriate place along a scale. Christine's question 6 could be presented as a continuum:

> How do you evaluate the significance of your dreams? Place an "X" on the continuum in the place that most closely reflects your view.
>
> |————|————|————|————|
>
> *My dreams always hold some meaning* — *My dreams are meaningless*

Though it is a bit more difficult to tabulate results of a continuum, this method often produces reliable answers if the instructions are clear.

Planning for Distribution

Surveys can be administered in person or by phone, with the surveyor asking each respondent the questions and recording the answers, or by letting respondents fill out the surveys themselves. Although there are some real advantages to administering the survey yourself (or lining up friends to help you do it), reflect on how much time you want to devote to gathering the information. How important will the survey be to your paper? Are the results crucial to your argument? If not, consider doing what Christine did: Print several hundred survey forms that are easy for respondents to fill out themselves, and distribute them with some help from your instructor or friends.

The next question is, How can you make sure you will get back as many of the survey forms as possible? That topic is more fully discussed in the third week. (See "Conducting Surveys," Chapter 3.) For now, draft a survey, and line up some people to review it for you, including your instructor.

LOOKING BACK BEFORE MOVING ON

With a tentative focus decided and a strategy planned, you're well prepared to immerse yourself in research. It might seem odd to do all this planning before you ever crack a card catalog, but it pays off in a much more efficient search.

Be prepared for surprises along the way. If the research and writing process suddenly leads you down trails you didn't expect, it doesn't mean you've done something wrong. If the thesis you declare this week is shot full of holes next week, it doesn't mean that you're stupid. Through research, you discover what you didn't know and what you didn't know you knew. Research is supposed to rattle your cage, particularly if you've chosen a topic about which you have some passion. Get ready for an adventure.

3

□ ■ □

The Third Week

WRITING IN THE MIDDLE

I was never crazy about taking notes for a research paper. Note-taking seemed so tedious. Instead, I developed a love affair with the photocopier and walked around sounding like a slot machine, my pockets full of change, ready to bolt to the nearest copier whenever I encountered a promising article. I collected these articles to read later. I also checked out scores of books that seemed useful, rather than taking the time to skim them in the library and jot down notes on what seemed important. I was quite a sight at the end of the day, walking back to my dormitory or apartment, reeling under the weight of a mound of books and articles, all precariously balanced, defying natural laws.

When the time came to begin writing my paper, the work seemed agonizingly slow. I would consult my meager notes, thumb through two or three books from the stack, reread a dog-eared copy of an article, stop and think, write a line or two, stop and go back to a book, and then maybe write another line or two. I was always a slow writer, but I now realize that one major reason I got bogged down writing my research paper drafts was my inattention to notetaking. I paid the price for doing so little writing before I had to do the writing.

I now believe that the writing that takes place in the *middle* of the research process—the notetaking stage—may be as important, if not more so, than the writing that takes place at the end—composing the draft. Writing in the middle helps you take possession of your sources and establish your presence in the draft. It sharpens your thinking about your topic. And it is the best cure for unintentional plaigirism.

I realize I have a sales job to do on this. Writing in the middle, particularly if you've been weaned on notecards, feels like busy work. "It gets in the way of doing the research," one student told me. "I just want to collect as much stuff as I can, as quickly as I can. Notetaking slows me down." Though it may seem inefficient, writing as you read may actually make your research *more* efficient. Skeptical? Read on.

Becoming an Activist Notetaker

Notetaking can and probably should begin the process of writing your paper. Notetaking is not simply a mechanical process of vacuuming up as much information as you can and depositing it on notecards or in a notebook with little thought. Your notes from sources are your first chance to *make sense* of the information you encounter, to make it your own. You do need more time to take thoughtful notes, but doing so pays off in writing a draft more quickly and in producing a paper that reflects your point of view much more strongly.

I'll show you what I mean. Here is a short passage from a book on soccer, *The Simplest Game,* by Paul Gardner:

> For American soccer the problem is two-fold. First the majority of people coaching youth soccer did not play the sport themselves, or played it only at a rather crude level. In other words, they are not in a position to coach by example. Second, even if the coaches were themselves skilled players, it is not at all clear to what extent ball skills can be coached.*

Jason, who was writing a paper that focused on the failings of American soccer, took the following notes on the passage. Notice how he used the source as a launching place for his own thinking. In this case, Jason didn't even quote Gardner but took several of the author's ideas and explored how they fit in his own experience.

> So many times when I was young and playing soccer the emphasis was placed simply on winning when it should have been placed on teaching the skills necessary to play. I remember being in the pack

*Paul Gardner, *The Simplest Game* (Boston: Little, Brown, 1976), 193.

> of kids chasing the ball around just kicking it. The coach I had never emphasized ball control, we kicked and ran. That's it. The fastest players and hardest kickers were the best. More time needs to be spent on handling the ball and knowing what to do with it in certain situations.*

Here is the passage in Jason's research paper that incorporated the earlier notes. Notice the similarity?

> So many times when I was young and playing soccer like Joey, the emphasis was on winning rather than the skills necessary to play. I remember being in the same pack of kids as Joey, chasing the ball around. The coach that I had never emphasized ball control. I can still hear him yell, "Dump and chase!" Although it is unclear to what extent ball skills can be coached, the coach should at least spend time with players showing them technique (Gardner 193).*

Because Jason was committed to thoughtful notetaking *as* he read, he got a jumpstart on his draft. But even more important, he took command of the material, using the original source purposefully, helping him convince readers that American soccer will never be competitive without major reforms.

Jason's complaint about the way young American soccer players are trained to "dump and chase" is actually a good metaphor for the way most students approach notetaking: Simply shovel as much information (mostly quotes) on notecards as you can and chase after more. But taking notes, like playing soccer, is a skill, and a remarkably undervalued one at that. Test your notetaking skills with the following exercise.

*Jason Pulsifer, University of New Hampshire, 1991. Used with permission.

❑ *EXERCISE 3.1*
Creative Translations

The following passage is from Andrew Merton's article "Return to Brotherhood."

> For many adolescent males just out of high school, the transition to college represents a first step in a struggle for a kind of "manhood" from which women are viewed as objects of conquest—worthy, but decidedly inferior, adversaries. The idea of women as equals is strange and inconvenient at best, terrifying at worst. Unfortunately, most colleges and universities provide refuges ideally suited to reinforce these prejudices: fraternities.*

In your notebook, rewrite the passage in your own words in roughly the same length. This is called *paraphrasing*. You'll find it's easier to do if you first focus on understanding what Merton is trying to say and then write without looking much at the passage, if possible. If this is an in-class exercise, exchange your rewrite with a partner. Then read the following section on plagiarism.

□ ■ □

Recognizing Plagiarism

Simply put, *plagiarism* is using others' ideas *or* words as if they were your own. The most egregious case is handing in someone else's work with your name on it. Some schools also consider using one paper to meet the requirements of two classes to be a grave offense. But most plagiarism is unintentional. I remember being guilty of plagiarism when writing a philosophy paper my freshman year in college. I committed the offense largely because I didn't know what plagiarism was and I hadn't been adequately schooled in good scholarship (which is no excuse).

The best antidote to plagiarism is good notetaking. Examine the following two student paraphrases of the Merton passage. If each appeared in a final paper as you see it here, which student do you think would be guilty of plagiarism?

STUDENT A

According to Merton, for a lot of adolescent males just out of high school, going to college is, among other things, a journey into "manhood."

*Andrew Merton, "Return to Brotherhood," *Ms.*, September 1985, 61+.

Women may become the respected enemy, seen as things to conquer. The notion that women might be equals is not a comfortable one for these men, who often find fraternities a refuge ideally suited to promoting this stereotype.

STUDENT B

Many young men graduating from high school come to the university on the threshold of "manhood." As they come to terms with this, women become "worthy, but decidedly inferior, adversaries," an attitude that makes the notion of equality between the genders "inconvenient" or even "terrifying." The college fraternity becomes a safe haven for these men (Merton 192).

Both students' versions follow the organization of the original material fairly closely. Although doing so is not necessarily a fault, it does increase the risk of plagiarism. Student A most clearly crossed that line. You may not have thought so at first. After all, the writer did mention Merton in the beginning, whereas the second student did not.

But a closer look will reveal a problem with version A: Merton's words have become the writer's. Notice the phrase "for a lot of adolescent males" in version A. Merton used nearly the same language. In the last line of the paraphrase, "ideally suited" and "refuge" have been borrowed directly from Merton, without giving him credit using quotation marks (even though the word order has been altered somewhat). There's another more serious problem with Student A's paraphrase: Though Merton was mentioned in the text of the paraphrase, he must still be cited as the source. Using the Modern Language Association (MLA) citation system (see Appendix A), a parenthetical reference should have been provided, listing the author and page number from which the material came.

Tactics for Avoiding Plagiarism

The use of quotation marks around the borrowed language would have helped Student A. Student B was careful about that; in her version, distinctive phrases from the original were all flagged by enclosing them in quotation marks. Though Student B didn't mention Merton, she did include a parenthetical citation of the source. But there's still a problem: Because the citation appears at the end of the paraphrase, it's unclear what part has been borrowed. Just the last line? Both paraphrases are poor, but it wouldn't take much to redeem Student B's version. She could simply clear up the confusion by mentioning Merton in the text. For example, she could begin as Student A did: "According to Andrew Merton, many young men . . ."

To avoid plagiarism, you must do two things: Let readers know (1) exactly what material is not your own and (2) exactly where that borrowed material came from. One notable exception to this is when you include what may be considered *common knowledge* in your paper. (See Chapter 4 for a fuller discussion of the common knowledge exception.) For example, you wouldn't need to cite the fact that John F. Kennedy was assassinated in Dallas in November 1963. That fairly well known piece of information doesn't belong to anyone, and it could be used without citation.

You'll learn more about how to cite material in the fourth week (Chapter 4), when you begin to write your paper. For now, keep the following in mind:

1. When you borrow language from an original source, always signal you've done so by using quotation marks.
2. Whenever possible, attribute the borrowed words, ideas, or information to its author *in your text* (e.g., "According to Andrew Merton" or "Merton argues" or "Merton observes," etc.).
3. Whether you attribute the borrowed material in your text or not, *you must cite* the original source. (For more information on how to use parenthetical citations, see "Citing Sources in Your Essay" in Appendix A.)

❑ *EXERCISE 3.2*
Checking for Plagiarism

Look over your partner's paraphrase (or your own if you're not in class) of the Merton passage from Exercise 3.1, and ask your partner to examine yours. Do either of you see anything that could be considered plagiarism? Pay particular attention to language. Are the word choices of the paraphrase distinct from those of the original passage? Are quotes used to signal borrowed language? Discuss ways to remedy any potential problems.

Don't be alarmed if you were guilty of plagiarism here. Your error was unintentional. You were also doing an exercise. If your paraphrase did have problems, think back on how you wrote it. Did you spend a lot of time rereading the original as you were writing? Unintentional plagiarism often occurs because you just can't free yourself from the author's words. That freedom comes, in part, by finding your own writing voice as you begin to take notes. Your voice will likely be quite different from those of your sources and will naturally lead you away from their language and toward your own.

□ ■ □

How to Be a Purposeful Notetaker

Paraphrasing

You've already had practice with the paraphrase (a restatement of a source in roughly equal length) in Exercise 3.2. You may have discovered that writing a good paraphrase is harder than you thought, especially when being cautious to avoid plagiarism. Paraphrasing requires that you take possession of the information, make sense of it for yourself, and then remake it sensibly in your own words. But paraphrasing gets easier with practice. And when you begin to master the art of the paraphrase, you will also begin to take charge of your topic, *using* information deliberately, according to your own purpose.

In the last exercise, you were asked to paraphrase a passage on how fraternities promote sexist attitudes among young men. Some of you might be young men, who also happen to be fraternity brothers. You may not have thought much of the author's point of view. Suppose you were writing a paper whose purpose was to argue *against* a movement to abolish fraternities. Would that alter how you paraphrased the passage? Would you recombine or emphasize different elements of it?

Melding Paraphrase and Purpose. Paraphrasing shouldn't be divorced from your purpose in writing the paper. Let me show you what I mean. The following line is from Paulo Friere's essay "The 'Banking' Concept of Education":

> In the banking concept of education, knowledge is a gift bestowed by those who consider themselves knowledgeable upon those whom they consider to know nothing.*

A fairly neutral paraphrase of the Friere passage might be something like this:

*Paulo Friere, "The 'Banking' Concept of Education," trans. Myra Bergman Ramos, in *Pedagogy of the Oppressed* (New York: Continuum, 1970).

> One approach to education is to view students as empty banks and teachers as depositors of knowledge.

But suppose you're taking notes for a paper that focuses on educational reform, and you really agree with Friere's view that the "banking concept" is suspect, seeing it as strong evidence that something is wrong with the system. Your paraphrase will likely be quite different, perhaps this:

> According to Paulo Friere, a noted critic of current teaching methods, all too often teachers see students as simply empty bank accounts into which they must deposit knowledge.

See the difference? In this paraphrase, the writer emphasizes who is making the claim—"a noted critic"—and also stresses the banking metaphor, which implies that teachers disparage students by considering them intellectually bankrupt. The writer not only captures Friere's theme but deliberately picks up on the tone of his passage. And why not? Friere's critical tone nicely serves the writer's purpose: arguing for educational reforms.

Let's take this one step further. If good notes reflect a writer who actively responds to what a source is saying, then a paraphrase (or a summary or quote) can be a launching place for the writer's own thinking, a tool for her own analysis. This is when notetaking becomes most useful. Look what the same writer might do with the Friere passage:

> Friere's "banking concept of education"--which suggests that students are things into which knowledge must be deposited--only further alienates them from learning, reinforcing the idea that knowledge isn't meant to be used but simply absorbed like a sponge.

Technically, this is still a paraphrase, but it's so much more than a parroting of information. The writer is taking Friere's idea and using it to make a point of her own, extending her own analysis. Friere steps in and out, while the writer takes over.

As you're taking notes, *look for these opportunities to actively use sources to push your own thinking.* Though such notetaking requires a little more thought, the notes you produce will later become the meat of your research paper. You are, in effect, writing the paper as you research, saving time later. Just one note of caution: In your enthusiasm, be careful not to misrepresent the original author's ideas.

You may think it will take forever to take notes like these, but, with practice, you will find notetaking goes more quickly than you thought. Again, the key is to find some distance from the source so you can reflect on what the author's saying and how it relates to what you think. Two things will help you get that distance, especially when paraphrasing: Write your notes in your own voice, and don't feel obligated to mimic the original author's organization and emphasis. Select, recombine, and use only what you think is important. Several techniques discussed later in this chapter, including "The Double-Entry Journal," also encourage this kind of notetaking. Give them a try.

Summarizing

I heard recently that in order to sell a movie to Hollywood, a screenwriter should be able to summarize what it's about in a sentence. "*The Big Chill* is a film about six friends from the sixties who reunite some years later and are suddenly confronted with reconciling past and present." That statement hardly does justice to the film—which is about so much more than that—but I think it basically captures the story and its central theme.

Obviously, that's what a *summary* is: a reduction of longer material into some brief statement that captures a basic idea, argument, or theme from the original. Like paraphrasing, summarizing often requires careful thought, since you're the one doing the distilling, especially if you're trying to capture the essence of a whole movie, article, or chapter that's fairly complex.

But many times, summarizing involves simply boiling down a passage—not the entire work—to its basic idea. Mark, who researched patterns of alcohol abuse among college students, found an article from the *Journal of Studies on Alcoholism* that surveyed college students on situations in which they might choose to intervene—or not to intervene—to help a fellow student with a drinking problem. Much of the article discussed methodology, which wasn't very useful to Mark's essay, but one passage had some interesting facts:

> Decisions not to intervene were reported less often by the respondents. In the drunk-driving situation, only 26% of the respondents in the sample (30% of the men and 24% of the women) indicated that, at least once, they both saw someone who was too drunk to drive and chose not to do or say anything. . . . Students indicated they chose not to intervene because they did not know the person well enough (24%) or that they were too drunk themselves to be of any help (24%). However, additional reasons for nonintervention in the drunk-driving situation included the anticipated response (17%) of the target (no matter what the intervener did, the behavior of the target could not be changed); the degree to which intervention would result in the intervenor's negative self-image (16%) (fear of looking foolish in front of others present); [and] misperceptions (13%) of the situation (unsure how drunk the target was). . . . Interestingly, men (23%) more frequently cited concern about creating a negative self-image than did women (11%), while women (20%) more frequently identified the anticipated response of the target as a reason for nonintervention than did men (14%).*

There's nothing particulary distinctive about how this information is presented (in fact, it's pretty dry), so there's nothing particularly quotable. And paraphrasing the whole passage seems unnecessary, since not all the information is important for Mark's purposes. But embedded in this passage is some pretty interesting stuff, especially given Mark's project: to explore how students' personal relationships influence their drinking habits.

A simple summary might be something like this:

> According to Thomas and Siebold, most college students say they would try to stop a drunk student from getting into his car and driving away. Roughly one-quarter of the students surveyed, however, wouldn't intervene, and 24 percent of those said it was because they were too drunk themselves.

This summary helps Mark emphasize only the information from this source that's important to his essay. He uses that information selectively and powerfully by putting it in his own words.

*Richard W. Thomas and David R. Siebold, "College Students' Decisions to Intervene in Alcohol-Related Situations," *Journal of Studies on Alcohol* 56.5 (1995): 580–588.

A summary such as this one, where you distill only part of a larger source, is selective. You choose to emphasize some key part of a source because it fits your paper's purpose. But the same warning applies to selective summarizing as was given earlier about paraphrasing: Don't misrepresent the general thrust of the author's ideas. Ask yourself, Does my selective use of this source seem to give it a spin the author didn't intend? Most of the time, I think you will discover the answer is no.

A summary, similar to a paraphrase, can prompt your own thinking. Remember Jason's notes on American soccer at the beginning of this chapter? His summary of a section of Gardner's book, which claimed that American soccer players are not taught necessary ball-handling skills, provided Jason with an opportunity to reflect on his own experience as a player and how that seemed to confirm Gardner's point.

As with paraphrases, your most useful summaries will be those that provide an opportunity for your own analysis, that get you thinking. As you take notes, look for chances to let your summaries take off. Explore connections and contradictions with other sources as well as with your own experiences and observations. For example, here's what Taylor, who is critical of fraternities, did in her notes with a summary of the Merton passage on fraternities, discussed earlier:

> Andrew Merton argues that fraternities are a "safe refuge" for maturing young men who are threatened by women. What makes them safe is what someone else called "a conspiracy of silence" between brothers about sexist behavior. It violates the code to confront a brother. An exclusively male culture will always treat women as outsiders.

Here, the writer's summary is the starting place for a brief discussion about Merton's premise, including a connection with something else Taylor had heard or read. Once again, you can see the writer taking over here, taking command of the original material, not just reciting its main point.

These kinds of notes, into which you inject your own thinking, may be new to you. This type of notetaking is a far cry from simply jotting down facts from the encyclopedia. But if you chose your topic because it's interesting to you, you may discover that it's quite natural to be an activist notetaker. You want to make sense of things because you want to learn.

Quoting

What will *not* seem new to you, most likely, is the habit of quoting sources in your notes and your paper. The quotation mark may be the student researcher's best friend, at least as demonstrated by how often papers are peppered by long quotes!

As a general rule, the college research paper should contain no more than 10 or 20 percent quoted material, but it's an easy rule to ignore. For one thing, quoting sources at the notetaking and drafting stages is quicker than restating material in your own words. When you quote, you don't have to think much about what you're reading; you just jot it down the way you see it and, if you have to, think about it later. That's the real problem with verbatim copying of source material: There isn't much thinking involved. As a result, the writer doesn't take possession of the information, shape it, and be shaped by it.

That's not to say that you should completely avoid quoting sources directly as a method of notetaking. If you're writing on a literary topic, for example, you may quote fairly extensively from the novel or poem you're examining. Or if your paper relies heavily on interviews, you'll want to bring in the voices of your subjects, verbatim.

When to Quote. As a rule, jot down a quote when someone says or writes something that is distinctive in a certain way and when restating it in your own words wouldn't possibly do the thought justice. I'll never forget a scene from the documentary *Shoah,* an eleven-hour film about the Holocaust, which presented an interview with the Polish engineer of one of the trains that took thousands of Jews to their deaths. Now an old man and still operating the same train, he was asked how he felt now about his role in World War II. He said quietly, "If you could lick my heart, it would poison you."

It would be difficult to restate the Polish engineer's comment in your own words. But more important, it would be stupid even to try. Some of the pain and regret and horror of that time in history is embedded in that one man's words. You may not come across such a distinctive quote as you read your sources this week, but be alert to *how* authors (and those quoted by authors) say things. Is the prose unusual, surprising, or memorable? Does the writer make a point in an interesting way? If so, jot it down.

Heidi, in a paper on the children's TV program *Sesame Street,* began by quoting a eulogy for Muppets creator Jim Henson. The quote is both memorable and touching. Heidi made an appropriate choice, establishing a tone that is consistent with her purpose: to respond to certain critics of the program. The fact that a quote sounds good isn't reason enough to use it. Like anything else, quotes should be used deliberately, with purpose.

There are several other reasons to quote a source as you're note-taking. Sometimes, it's desirable to quote an expert on your topic who is widely recognized in the field. Used to support or develop your own assertions, the voice of an authority can lend credit to your argument and demonstrate your effort to bring recognized voices into the discussion.

Another reason to quote a source is when his explanation of a process or idea is especially clear. Such quotes often feature metaphors. Robert Bly's *Iron John,* a book that looks at American men and their difficult journey into manhood, is filled with clear and compelling explanations of that process. As a son of an alcoholic father, I found Bly's discussion often hit home. Here, using a metaphor, he explains in a simple but compelling way how children in troubled homes become emotionally unprotected, something that often haunts them the rest of their lives:

> When a boy grows up in a "dysfunctional" family (perhaps there is no other kind of family), his interior warriors will be killed off early. Warriors, mythologically, lift their swords to defend the king. The King in a child stands for and stands up for the child's mood. But when we are children our mood gets easily overrun and swept over in the messed-up family by the more powerful, more dominant, more terrifying mood of the parent. We can say that when the warriors inside cannot protect our mood from being disintegrated, or defend our body from invasion, the warriors collapse, go into a trance, or die.*

I'm sure there's a more technical explanation for the ways parents in dysfunctional families can dominate the emotional lives of their children. But the warrior metaphor is so simple, and that is, partly, its power. As you read or take notes during an interview, be alert to sources or subjects who say something that gets right to the heart of an important idea. Listen for it.

If your paper is on a literary topic—involving novels, stories, poems, and other works—then purposeful and selective quoting is especially important and appropriate. The texts and the actual language the writers use in them are often central to the argument you're making. If you're writing about the misfit hero in J. D. Salinger's novels, asserting that he embodies the author's own character, then you'll have to dip freely into his books, quoting passages that illustrate that idea. (See Appendix C for two essays on literary topics that use quotes effectively.)

*Robert Bly, *Iron John: A Book about Men* (Reading, MA: Addison-Wesley, 1990), 147.

Quoting Fairly. If you do choose to quote from a source, be careful to do three things: (1) Quote accurately, (2) make sure it's clear in your notes that what you're jotting down is quoted material, and (3) beware of distorting a quote by using it out of context. The first two guidelines protect you from plagiarism, and the last ensures that you're fair to your sources.

To guarantee the accuracy of a quote, you may want to photocopy the page or article with the borrowed material. A tape recorder can help in an interview, and so can asking your subject to repeat something that seems especially important. To alert yourself to which part of your notes is a quote of the source's words, try using oversized quotation marks around the passage so that it can't be missed.

Guarding against out-of-context quotations can be a little more difficult. After all, an isolated quote has already been removed from the context of the many other things a subject has said. That shouldn't be a problem if you have represented her ideas accurately. However, sometimes a quote can misrepresent a source by what is omitted. Simply be fair to the author by noting any important qualifications she may make to something said or written, and render her ideas as completely as possible in your paper.

❑ *EXERCISE 3.3*

Good Notes on Bad Writing

Get some practice with purposeful notetaking with the article "The Importance of Writing Badly," which follows. First, read it carefully, paying attention to what you agree or disagree with in the essay. Then, in your research notebook:

1. Summarize the article in a sentence or two.
2. Select at least two passages that seem quotable to you, and carefully jot them down.
3. Below your summary, do a five-minute fastwrite in which you explore your reaction to the article. What did it make you think or remember about the way you learned to write? What assertions did you really agree with or disagree with? Why? What is the author missing in his argument? How might you go further?
4. Now take ten minutes and compose a paragraph that brings in summary, quote, and perhaps paraphrase to support your own reactions to the ideas in the article. Obviously, take care to avoid any plagiarism by using your own voice, by attributing the source, and by using quotation marks to signal borrowed language. Use the article, but make it your own.

□ ■ □

The Importance of Writing Badly

by Bruce Ballenger

I WAS grading papers in the waiting room of my doctor's office the other day, and he said, "It must be pretty eye-opening reading that stuff. Can you believe those students had four years of high school and still can't write?"

I've heard that before. I hear it almost every time I tell a stranger that I teach writing at a university.

I also hear it from colleagues brandishing red pens who hover over their students' papers like Huey helicopters waiting to flush the enemy from the tall grass, waiting for a comma splice or a vague pronoun reference or a misspelled word to break cover.

And I heard it this morning from the commentator on my public radio station who publishes snickering books about how students abuse the sacred language.

I have another problem: getting students to write badly.

Most of us have lurking in our past some high priest of good grammar whose angry scribbling occupied the margins of our papers. Mine was Mrs. O'Neill, an eighth grade teacher with a good heart but no patience for the bad sentence. Her favorite comment on my writing was "awk," which now sounds to me like the grunt of a large bird, but back then meant "awkward." She didn't think much of my sentences.

I find some people who reminisce fondly about their own Mrs. O'Neill, usually an English teacher who terrorized them into worshipping the error-free sentence. In some cases that terror paid off when it was finally transformed into an appreciation for the music a well-made sentence can make.

But it didn't work that way with me. I was driven into silence, losing faith that I could ever pick up the pen without breaking the rules or drawing another "awk" from a doubting reader. For years I wrote only when forced to, and when I did it was never good enough.

Many of my students come to me similarly voiceless, dreading the first writing assignment because they mistakenly believe that how they say it matters more than discovering what they have to say.

The night before the essay is due they pace their rooms like expectant fathers, waiting to deliver that perfect beginning. They wait and they wait and they wait. It's no wonder the waiting often turns to hating what they have written when they finally get it down. Many pledge to steer clear of my English classes, or any class that demands much writing.

My doctor would say my students' failure to make words march down the page with military precision is another example of a failed educational system. The criticism sometimes takes on political overtones. On my campus, for example, the right-wing student newspaper demanded an entire semester of Freshman English be devoted to teaching students the rules of punctuation.

There is, I think, a hint of elitism among those who are so quick to decry the sorry state of the sentence in the hands of student writers. A colleague of mine, an Ivy League graduate, is among the self-appointed grammar police, complaining often about the dumb mistakes his students make in their papers. I don't remember him ever talking about what his students are trying to say in those papers. I have a feeling he's really not all that interested.

Concise, clear writing matters, of course, and I have a responsibility to demand it from my students. But first I am far more interested in encouraging thinking than error-free sentences. That's where bad writing comes in.

When I give my students permission to write badly, to suspend their compulsive need to find the "perfect way of saying it," often something miraculous happens: Words that used to trickle forth come gushing to the page. The students quickly find their voices again, and even more important, they are surprised by what they have to say. They can worry later about fixing awkward sentences. First they need to make a mess.

It's harder to write badly than you might think. Haunted by their own Mrs. O'Neill, some students can't overlook the sloppiness of their sentences or their lack of eloquence, and quickly stall out and stop writing. When the writing stops, so does the thinking.

The greatest reward in allowing students to write badly is that they learn that language can lead them to meaning, that words can be a means for finding out what they didn't know they knew. It usually happens when the words rush to the page, however awkwardly.

I don't mean to excuse bad grammar. But I cringe at conservative educational reformers who believe writing instruction should return to primarily teaching how to punctuate a sentence and use *Roget's Thesaurus*. If policing student papers for mistakes means alienating young writers from the language we expect them to master, then the exercise is self-defeating.

It is more important to allow students to first experience how language can be a vehicle for discovering how they see the world. And what matters in this journey—at least initially—is not what kind of car you're driving, but where you end up.

Three Notetaking Tips That Will Save You a Headache Later On

1. *Before you take any notes, write down the complete bibliographic information on the source (author, title, publication information).* Why? Because you'll save yourself a late-night trip back to the library to get the bibliographic information on that one source you forgot to write down. In the first week, the library and Internet exercises (Exercises 1.4 and 1.5) gave you some practice in getting what you need to cite a source. You'll need to cite slightly different information for a journal article than for, say, a book. Get all of the information for every source. (See "Preparing the 'Works Cited' Page" in Appendix A for exactly what bibliographic information you'll need to list different kinds of sources if you're using MLA format in your paper.)

2. *When you do take notes, make sure you get the page the material was taken from.* It's so easy to forget. But as you'll see later, you need to show exactly where you borrowed from a source when you cite the material in the text of your paper. If you go to the next page of a source in the middle of your notes on it, use the symbol "/ " to remind yourself of that fact. If you skip more than one page, use "// " in your notes as a reminder. Or use a system like the double-entry journal (see later in this chapter) that encourages you to note page numbers in the margins as you go along.

3. *Be accurate.* Sure, you think. That's obvious. But there are so many ways to mess up. The most common mistake is to quote a source inaccurately—inadvertently leaving out a word, using the wrong word, misspelling a word, or leaving out a chunk of the original passage without signaling to your readers that you've done so. Remember to use an *ellipsis* (three dots, ". . .") to indicate where you've left out a portion of the author's text that you decided wasn't important to your point. Also keep track of when you're quoting in your notes by using oversized quotation marks that cannot be missed. Most important, make sure you've summarized the author's ideas fairly and accurately and that he meant to say what you've represented as such.

NOTETAKING TECHNIQUES

In the previous edition of *The Curious Researcher,* I confessed to a dislike of notecards. Apparently, I'm not the only one. Mention notecards, and students often tell horror stories. It's a little like talking about who has the most horrendous scar, a discussion that can prompt participants to expose knees and bare abdomens in public places. One student even mailed me her notecards—fifty bibliography cards and fifty-three notecards, all bound by a metal ring and color coded. She assured me that she didn't want them back—ever. Another student told me she was required to write twenty notecards a day: "If you spelled something wrong or if you put your name on the left side of the notecard rather than the right, your notecards were torn up and you had to do them over."

It is true, of course, that some students find recording information on notecards an enormously useful way of organizing information. And some teachers have realized that it's pretty silly to turn notetaking into a form that must be done "correctly" or not at all. For these reasons, I included suggestions about how to use notecards effectively in the previous edition. But in good conscience, I can't do it anymore. I no longer believe that 3" × 4" or 4" × 6" cards are large enough to accommodate the frequently messy and occasionally extended writing that often characterizes genuinely useful notes. Little cards get in the way of having a good conversation with your sources.

There are a range of alternative notetaking strategies that do promote having such a conversation through writing. I recommend two of them here: the double-entry journal and the research log. These methods are not meant to be strictly prescriptive; each can be customized to suit your own sensibilities. A few years from now, by the time you've written a score of research papers for college courses, you'll have refined your own notetaking method. In the meantime, try out these approaches and attempt to make them your own.

The Double-Entry Journal

The double-entry approach (see Figure 3.1) is basically this: Divide each page of your research notebook into two columns (or use opposing pages). On each left side, compile your notes from a source—paraphrases, summaries, quotes—and on each right side, comment on them. Your commentary can be pretty open ended: What strikes you? What was confusing? What was surprising? How does the information stand up to your own experiences and observations? Does it support or contradict your thesis (if you have one at this point)? How might you use the information in your paper? What purpose might it serve? What do you think of the source? What further questions does the information raise that might be worth investigating? How does the information connect to other sources you've read?

There are a variety of ways to approach the double-entry journal. If you're taking notes on a photocopied article or a book you own, try reading the material first and underlining passages that seem important. Then, when you're done, transfer some of that underlined material—quotes, summaries, or paraphrases—into the left column of your journal. Otherwise, take notes in the left column *as* you read.

Notes from Source

- In this column, collect direct quotations, paraphrases, and summaries of key ideas that you cull from your source.
- Collect material that's relevant to your project, but also write down passages, facts, and claims from the source that you find surprising or puzzling or that generate some kind of emotional response in you.
- Make sure you write down this material carefully and accurately.
- Don't forget to include the page number from the source to the left of the borrowed material or idea.

Fastwrite Response

- In this column, think through writing about some of the information you collected in the other column. This will likely be a messy fastwrite but a focused one.
- Whenever your writing dies, skip a space, look to the left, and find something else to respond to.
- Some questions to ponder as you're writing might include:
  1. What strikes me about this?
  2. What are my first thoughts when I consider this? And then what? And then? And then?
  3. What exactly does this make me think of or remember?
  4. How would I qualify or challenge this author's claim? In what ways do I agree with it?
  5. What else have I read that connects with this?
  6. How do I feel about this?

FIGURE 3.1 Double-entry journal method

When you've finished taking notes, it's time to do some exploratory writing in the right column. This territory belongs to you. Here, through language, your mind and heart assert themselves over the source material. Use your notes in the left column as a trigger for writing in the right. Whenever your writing stalls, look to the left. The process is a little like watching tennis—look left, then right, then left, then right. Direct your attention to what the source says and then to what *you* have to say about the source. Keep up a dialogue.

Figures 3.2 and 3.3 illustrate how the double-entry journal works in practice. Note these few features that are common to both approaches:

- Bibliographic information is recorded at the top of the page. Do that first, and make sure it's complete.

- Page numbers are included in the far-left margin, right next to the information that was taken from that page. Make sure you keep up with this as you write.

Ehrenstein, David. "Film in the Age of Video." Film Quarterly 49.3 (1996): 38-42.

38 ". . . today the once distinct spheres of theatrical and home exhibition have been radically conflated."	*Let's see. So the "spheres" of showing films in theaters and in home video have been "conflated." What I think he means is that movies are now produced with both means of showing them in mind, which would seem to have implications for how they're made these days. E. talks later in the article about this, I think, when he mentions how only the dimension of sound has been preserved from the old days of big screens in dark theaters. The "big image" is lost. I'm not sure what this means, exactly.*
39 That's Entertainment 3, which started as video is "less a spectacular to be enjoyed in a darkened theater than a work of historical and cultural research that invites detailed analysis—a mode of consumption that home video, by its very nature, encourages." *40 "There are any number of (video) sets devoted to films, old and new, that enable the average everyday consumer to examine cinema now as never before."* *"The illusion of depth" is destroyed on home video.*	*I like this phrase that home video represents a particular "mode of consumption" for film. You may see more than one film at a sitting, in a lighted room, and you can rewind and reexamine favorite scenes and images. There is something about sitting in the dark, too, watching a big screen with a few hundred other people. It's like you experience nothing but image because there's nothing else to see. And somehow the act of watching with strangers, instead of in your living room by yourself or with friends, creates a kind of community. But E. talks here about even the theater experience is no longer "distinctive." But he doesn't really say why. What exactly was lost with the*

FIGURE 3.2 Sample double-entry journal. Here, the writer concentrates on collecting and responding to quoted material from the source.

41 "To remember the movie palaces, with enormous images floating in a velvety darkness, framed by curtains that never seemed to close on an ever-shifting program (features, cartoons, shorts, news, coming attractions) is not to indulge in nostalgia but rather to note how radically cinematic object relations have changed. There's nothing in any way distinctive about the modern theatrical movie-going experieence, save the sound.	*disappearance of the "movie palace?" Size, for one thing. Maybe that's one way that home video and movies in theater have been "conflated." Because, as E. says, films are now made for both video and theaters, and the "illusion of depth" as well as size of the image is destroyed by home video, then there's really no need for the really big screen of the "movie palace." Instead, we now have theaters divided and subdivided into eight theaters. Screens have shrunk, rooms have shrunk, and the theater experience begins to approximate home video. That leads to this "era of lowered cinematic expectations."*
42 "We have entered an era of lowered cinematic expectations."	
Jacque Rivette: "The cinema is necessarily fascination and rape, that is how it acts on people; it is something pretty unclear, something one sees shrouded in darkness."	*I don't get this quote at all. How can Rivette compare the experience of watching a film as both "fascination and rape?" Does he mean that film does violence in the dark to viewers in the same way a rapist would?*
	Check Rivette cite.

FIGURE 3.2 Continued

■ While the material from the source in the left column may be quite formal or technical—as it is in Figure 3.3—the response in the right column should be informal, conversational. Try to write in your own voice. Find your own way to say things. And don't hesitate to use the first person: *I*.

■ As you read the writers' responses to their sources in these two examples, notice how often the writers use their own writing to try to question a source's claim or understand better what that claim might be (e.g., "What the authors seem to be missing here . . ." and "I don't get this quote at all . . .").

Fredrick, Christina M. and Virginia M. Grow. "A Mediational Model of Autonomy, Self-Esteem, and Eating Disordered Attitudes and Behaviors." Psychology of Women Quarterly 20 (1996): 217-228.

218 Study begins by reviewing evidence that eating disorders and "autonomy" are closely related. "Wagner et. al found . . . that eating disordered individuals not only perceive themselves as lacking independence, but that this lack of self-determination extends into social situations, such as dating and attending parties.

At first the whole premise of this study seemed totally obvious. Of course women who have low self-esteem and don't feel independent are going to be at risk for eating disorders and maybe any number of other problems. But then I began to see what they were getting at: that there might be some kind of "developmental" thing going on here. That lack of autonomy might lead to low self-esteem and that might lead to eating disorders. Give girls a strong feeling of independence then and you're likely to help them avoid eating disorder.

Low self-esteem has also been shown to be linked to eating disorders.

219 Authors believe that autonomy and self-esteem may have a relationship to each other: "We would predict that by the time people reach adulthood, they would have internalized messages from their environment pertaining to their needs for autonomy with ramifications for their self esteem. . . . For women in Western societies, one way to cope with intrapsyhic distress from unfulfilled needs for autonomy and resulting low self-esteem may be to focus on body-related issues."

Why would "body-related issues" naturally develop from lack of autonomy and low self-esteem? The key, I think, is that autonomy is all about control. The one thing an anorexic or bulimic thinks she can control is her body.

FIGURE 3.3 Sample double-entry journal. This writer, taking notes on an article from a scholarly journal, collects and responds to summaries, quotes, and paraphrases from the source.

220 Authors describe a developmental pathway: lack of autonomy leads to low self-esteem leads to eating disorders.	
Studied 71 women in an undergrad psych class at Univ. of Rochester.	
224 ". . . autonomy provides a fundamental foundation for true self-esteem and the development of other healthy personality characteristics."	This seems like a really important idea.
". . . a young woman who grows up in an environment that fails to support her needs for autonomy may learn to shift her focus from satisfying her own needs to satisfying the needs of others. Thus, she may believe that she is valued not as a separate worthwhile individual but rather for what she does to make others happy. . . . By learning to be 'perfect' and seemingly in control, the girl subsequently fails to develop a separate sense of self-esteem because her feelings of worth are contingent on pleasing others."	The connection I can't quite see here is why unfulfilled needs for autonomy would lead someone to focus on pleasing others. What the authors seems to be missing here is the connection between autonomy and control and eating disorders. Like what I was saying earlier: If you don't feel you have control over your life, you can exert control over your body (or so you think). "The effort to be perfect" is a set-up for failure, and this leads to low self-esteem. But I'm still not sure this low self-esteem is because she failed others. It's because she failed herself. That's my experience with it.
Study concludes that there is a "statistically suggestive . . . developmental pattern" of lack of autonomy leading to low self-esteem leading to eating disorders. Treatment strategy? Deal with lack of autonomy.	

FIGURE 3.3 Continued

- Seize a phrase from your source, and play out its implications; think about how it pushes your own thinking or relates to your thesis. For example, the student writing about the rise of home video (Figure 3.2) plays with the phrase "mode of consumption"—a particular way of using film that the author believes home video encourages—and she really takes off on it. It leads her to a meditation on what it's like to see movies in theaters and what might be lost in the transition to home viewing.

- Use questions to keep you writing and thinking. In both examples, the writers frequently pause to ask themselves questions—not only about what the authors of the original sources might be saying but what the writers are saying to themselves as they write.

What I like about the double-entry journal system is that it turns me into a really active reader as I'm taking notes for my essay. That blank column on the right, like the whirring of my computer right now, impatiently urges me to figure out what I think through writing. All along, I've said the key to writing a strong research paper is *making the information your own.* Developing your own thinking about the information you collect, as you go along, is one way to do that. Thoughtful notes are so easy to neglect in your mad rush to simply take down a lot of information. The double-entry journal won't let you neglect your own thinking, or at least, it will remind you when you do.

The double-entry system does have a drawback. Unlike index card systems, double-entry journals don't organize your information particularly well. A lot of page flipping is involved to find pieces of information as you draft your paper. But I find I often remember which sources have what information, partly because I thought about what might be important as I read and took notes on each source.

The Research Log

The double-entry journal promotes a conversation between the researcher and her source. First, the source speaks in the left column through quotation, paraphrase, or summary, and then, the writer responds in the right column. In theory, this dialogue continues as the writer keeps traversing the boundary between the source's information and what she notices about it—that is, between an author's claims and the writer's own understanding of they might mean. This process of working things out through writing is mediated by the writer's own language; as she finds her own words to express her understanding of what she's read, she begins to assert control over her sources, rather than the other way around.

The research log described in this section (see Figure 3.4) promotes a similar dialogue but with a few differences. One is that the researcher starts with a monologue and always gets the last word—the Jay Leno approach. Another difference is that the research log may be more adaptable than the double-entry journal for researchers who prefer to write on computers. The standard format of the research log can serve as a template, which can be retrieved whenever you're ready to take notes on another source. Those notes can then be easily dropped into the draft as needed, using the "Cut-and-Paste" feature of your word-processing program. Obviously, the research log format can also be used in your notebook, if you're writing notes by hand.

The first entry in a research log should be a brief statement of the topic—perhaps a working title (see "Project" in Figure 3.4). Next, jot down basic bibliographic information ("Citation"), preferably written in MLA format so if you're composing on a computer, you can later cut and paste it onto the "Works Cited" page. Also note the date on which you entered this information in your research log.

The next major entry is titled "What Strikes Me Most." After you've carefully read the source once or twice and marked it up, set aside seven minutes (or more, if you really get going) to fastwrite your reaction to the piece. This is an open-ended response, but consider the following questions as prompts:

- What strikes me as the most important thing the author was trying to say?
- What surprised me most?
- What do I remember best?
- How did it make me feel?
- What seemed most convincing? Least convincing?
- How has it changed my thinking on my topic?
- How does it compare to other things I've read?
- What other research possibilities does it suggest?

As you write, resist the temptation to pick up the source and look it over again. This is a stand-up routine; you're on your own.

Your response can be personal or analytical or both. In Figure 3.4, the writer of the sample research log is interested in why frogs and other amphibians have disappeared from lakes and ponds in recent years. His original source was a three-page Internet document, an article that first appeared in the *San Francisco Chronicle.* (Entries on longer sources will likely be more extensive than the sample log entry illustrated here.) Notice that what struck this writer covered a variety of topics: how the piece made him feel, points that he hadn't encountered elsewhere, attempts to reconstruct some potentially useful facts, and the beginning of an anecdote about a childhood pond full of frogs.

Project: The Disappearing Frog

Citation: Petit, Charles. "Disappearance of toads, frogs has some scientists worried." San Francisco Chronicle 20 April 1992: 3 pgs. NEWS WEEK2. Online. Prodigy. 12 August 1996. Available HTTP: http://www.cs.yale.edu/homes/sjl/froggy/frogs-disappear.txt

Date: 8/12/96

What Strikes Me Most:
I was relieved to read in this article that at least a few biologists believe that the trend of frog disappearance around the world may not be terribly unusual but rather a natural cycle of growth and decline in numbers. But clearly that one expert from Rutgers that made that claim in this article was in the minority. I need to follow up and find if he has said more in any other articles. The key thing in this article is the point that the frog decline appears to be worldwide, so the cause is probably global. The article cited at least three possiblities: UV light from the ozone layer depletion, toxic substances, and one other that I can't remember. It mentioned a five-year study to determine which might be the cause. Another thing that struck me was the tone of some of the biologists' comments about the problem. One called the situation a "catastrophe" already. If he's right, then a five-year study may simply confirm that. It might be too late to do anything about it. I keep thinking as I read an article like this of the pond at the Wisconsin farm I spent time at as a child going silent.

Source Notes:
Article attributes decline to three possible causes:
1. acid rain
2. "thinning ozone layer"
3. wind-blown toxics (2)

Peter Morin, Rutgers University, called disappearance of amphibians "a phenomenon with much empirical support, not much data." Predicts that if researchers looked hard enough, they just might find that "there are places right around the corner that are full of frogs." (2)

FIGURE 3.4 Sample research log

"These are not just declines but appear to be absolute losses," said John Wright, who curates herpetology at the Los Angeles Museum of Natural History. "They are not just dropping down, they are in catastrophe." (3)

Study of back country in Cascades looked at 50 areas, including ponds where frogs typically found. "We found two," said Gary Fellers, a biologist. "Not frogs in two places. Two frogs total." (1)

"Scientists are hard pressed to understand how a diverse order of animals that has been on Earth for 200 million years should be highly vulnerable to an environmental change so subtle that experts cannot agree what it is." (2)

The Source Reconsidered:
I'm a little mystified by that last quote. Why is it surprising that even a creature that is a 200-million-year survivor should be vulnerable to subtle enviornmental changes, particularly if historically they're relatively recent human-induced changes? I wonder if there are certain qualities of so-called indicator species that the frog doesn't meet. Is that why scientists would be surprised by how easily frogs are affected by environmental change? Need to talk to somebody about that. My reconsideration of the article has deepened my sense that, notwithstanding the statement of the Rutgers guy, it's highly likely that the disappearance globally of amphibians is not part of a natural variation in numbers of the creatures. The Cascades study was alarming. If, as Morin suggests, they look hard enough they might "find places full of frogs," doesn't it seem odd that a look in 50 places only produced 2?

FIGURE 3.4 Continued

Next, mine the source for nuggets. Return to it, and reread those sections that seemed important. Under "Source Notes," collect quotes, summaries, paraphrases, facts—some of the things you underlined or circled or flagged that seemed important or interesting. Make sure you include the page number at the end of each item you've extracted from the source. Notice in the sample research log that the writer mined several interesting quotations from the frog article, as well as a factual clarification of the causes of amphibian disappearance, a point of confusion in the earlier fastwrite.

Finish the entry on each source with another fastwrite. Under the heading "The Source Reconsidered," think again through writing about how the article or book has altered your understanding of your topic. You'll probably find yourself writing about some of the nuggets you mined—how you now understand them, how they're significant to your project, or what new questions they raise. That's exactly the direction the writing took in the sample research log. The writer found himself seizing on one of the quotes, raising a question about it, and then speculating about possible answers. If you get stuck during the fastwrite, return to the "Source Notes" section and find some information that might be worth thinking about through writing. You might also use this prompt to either launch you or keep you going: "As I reconsider what this source said, I now see . . ." Write as much as you can.

This is only one of the many possible formats a research log might take. But the process of creating a log is essentially this: The writer reads a source, looks away to reflect on what he has read, has another direct encounter with the source, and then looks away again. Unlike the double-entry journal, the research log suggested here places more emphasis on the looking away, the process of working things out in conversation with oneself. Your research log can become a kind of narrative—a narrative of thought—in which you tell yourself the story of your evolving understanding of what you've read.

DIGGING DEEPER FOR INFORMATION

At the end of the third week of the research assignment last semester, Laura showed up at my office, looking pale.

"I spent all night at the library, and I couldn't find much on my topic," she said. "What I *could* find, the library didn't have—it was missing, or checked out, or wasn't even part of the collection. I may have to change my topic."

"I hate libraries!" she said, the color returning to her face.

Laura's complaint is one that I hear often at this point in the research process, especially from students who have dutifully tried to find a narrow focus for their papers, only to realize—they think—that there isn't enough information to make the topic work. They have tried the card catalog, and InfoTrac, and the newspaper indexes. They have even tried the *Social Science Index* or *Humanities Index* or several CD databases for journal articles. The students found a few articles but not enough for a ten-page paper. Like Laura, they may decide to broaden their focus or bail out on their topic altogether, even though they're still interested in it.

I always give these frustrated students the same advice: Don't despair yet. And don't give up on your narrow focus or your topic until you've dug more deeply for information. There are still some more specialized indexes to try and some nonlibrary sources to consider. You are, in a sense, like the archaeologist who carefully removes the dirt from each layer of a dig site, looking to see what it might reveal. If little turns up, the next layer is systematically explored and then the next, until the archaeologist is convinced she's digging in the wrong place. Student researchers too often give up the dig before they've removed enough dirt, believing too quickly there's nothing there.

If you're still curious about your topic and your tentative focus but you're not finding much information, work through the following three levels of the search before you decide to explore different ground. It might also be productive to expand the site of your search; you might basically be looking in the right place but not ranging far enough, perhaps limiting yourself to looking at books and articles when the real riches are in less conventional sources, possibly outside the library.

To help keep track of the ground you cover as you dig more deeply into your topic, check off the sources you've consulted at the end of each section in "Mapping Your Search," listing the names of the sources you used in the blanks. Your instructor may ask you to show her the trails you've followed as you discuss where else to look.

First-Level Searching

The library exercise completed in the first week (Exercise 1.4) asked you to search basic reference sources: general encyclopedias, books, periodicals, newspapers, and government documents. Consider these sources the first level of the dig. Make sure that you check *all* of them before you move to the next level. Here are a few sources that students commonly neglect:

- *Bibliographies in the backs of articles and books.* These can provide a wealth of possibilities. Look for additional sources cited by the author that seem promising.

- Social Science, Humanities, Art, *or* General Science Indexes *(books or computer databases).* Remember, these are general indexes to journal articles published every year. Go back a few years if you haven't done so already. Because these are less familiar indexes, students often neglect to use them. It's crucial to at least check these for articles on your topic.

- *Government documents.* This is a much neglected source of information from the largest publisher in the world. Make sure you check the *Monthly Catalog of U.S. Government Publications.*

- *Alternative subject headings.* Maybe you've been too wedded to a particular subject heading in doing your initial search. Try some variations. Go back to the *LC Subject Headings* for some other possibilities, or pay attention to the "see also" prompts in the indexes you're looking at. (See "The Story of a Search" in Chapter 1 for inspiration.)

MAPPING YOUR SEARCH:
THE FIRST LEVEL

Check off the sources you've consulted, and list the names of indexes you've reviewed in the spaces provided.

☐ General encyclopedias ______________________________

____

☐ Card catalog

☐ Popular periodical indexes ______________________________

____

☐ Academic journal indexes ______________________________

____

☐ Newspaper indexes ______________________________

____

☐ Government document indexes ______________________________

____

Second-Level Searching

By the time you reach this level in your dig, you will have at least consulted the major general indexes on your topic. If you're swamped with information at this point, it means one of two things: You're a thorough researcher, or your focus is too broad. You decide which it is, though I strongly encourage you to consider tightening your focus and resuming the dig. You're likely to find even more interesting stuff that will contribute to a paper that will probably surprise both you and your readers.

If you're still coming up empty handed and not finding much information after completing the first-level search, obviously, proceed with the second level.

For all of the searches suggested in the following sections (except those of computer databases), Balay's *Guide to Reference Books* is invaluable.* It's like a mall—you can do all your shopping there. All you need to know is where to look, and to determine that, place your topic in one or more general fields: humanities, social and behavioral sciences, history and area studies, or science, technology, and medicine. Under each general topic, find several more specific areas of study that might include your topic. Look for the specialized reference sources in that field that seem promising.

Guide to Reference Books is my favorite general reference work of its type, but if your library doesn't have it, try these alternatives:

Hillard, James. *Where to Find What: A Handbook to Reference Service.* Metuchen, NJ: Scarecrow, 1984.

McCormick, Mona. *The New York Times Guide to Reference Materials.* New York: Times Books, 1985.

Specialized Indexes to Journals

You're familiar with the *Social Science* and *Humanities Indexes,* which are general indexes to journal articles in various fields. What you may not know is that each discipline has its own more specialized indexes. For example, in psychology, *Psychological Abstracts* provides brief summaries of research published in the field every year in all the key journals. (The CD version is called *Psyclit.*) In political science, there's the *Public Affairs Information Service* (*PAIS*), often available as a computer database. Philosophers have the *Philosopher's Index,* and biologists have *Biological Abstracts* (also on CD). Often, these specialized indexes include material that you won't find in the more general indexes because specialized indexes cover more journals in a given field. That's why they're worth checking, even if you've been through the *Social Science, General Science,* or *Humanities Indexes.*

Following is a list of one or two key indexes in each field. There are more. For a more complete list, check the *Guide to Reference Books.* You can find the indexes listed most often as bound volumes in your campus library, but most are also available as databases on CD. Ask your reference librarian in which forms they are available.

*Robert Balay, ed., *Guide to Reference Books,* 11th ed. (Chicago: ALA, 1996). (Edited by Eugene Sheehy in earlier editions.)

ART

Art Index. New York: Bowker, 1929–date.

BIOLOGICAL SCIENCES

Biological Abstracts. Philadelphia: Biological Abstracts, 1926–date.

Biological and Agricultural Index (electronic source).

BUSINESS

Business Periodicals Index. New York: Wilson, 1958–date.

ABI/INFORM (electronic source).

CHEMISTRY

Chemical Abstracts: Key to the World's Chemical Literature. Columbus, OH: American Chemical Society, 1907–date (weekly).

COMMUNICATIONS

Communications Abstracts. Beverly Hills, CA: Sage, 1978–date.

COMPUTER SCIENCE

Applied Science and Technology Index. New York: Wilson, 1958–date.

ENVIRONMENT

Biological Abstracts. Philadelphia: Biological Abstracts, 1926–date.

ECONOMICS

Journal of Economic Literature. Nashville, TN: American Economic Association, 1964–date.

Economic Literature Index (electronic source).

EDUCATION

Education Index. New York: Wilson, 1929–date.

ERIC (electronic source).

ENGLISH AND LITERATURE

MLA International Bibliography of Books and Articles on the Modern Language and Literatures. New York: Modern Language Association, 1921–date.

FILM

Film Literature Index. Albany, NY: Film and Television Documentation Center, 1973–date.

GEOGRAPHY

Social Science Index. New York: Wilson, 1974–date.

GEOLOGY

Bibliography and Index of Geology. Boulder, CO: AGA, 1933–date.

Applied Science and Technology Index (electronic source).

HEALTH AND PHYSICAL EDUCATION

Education Index. New York: Wilson, 1929–date.

Medline (electronic source).

HISTORY

America: History and Life. Santa Barbara, CA: ABC-Clio, 1964–date.

JOURNALISM

Humanities Index. New York: Wilson, 1974–date.

Magazine Index (electronic source).

MATHEMATICS

General Science Index. New York: Wilson, 1978–date.

MEDICINE

Cumulated Index Medicus. Bethesda, MD: U.S. Department of Health and Human Services, 1959–date.

Medline (electronic source).

MUSIC

Music Index. Warren, MI: Information Coordinations, 1949–date.

PHILOSOPHY

Philosopher's Index. Bowling Green, KY: Bowling Green University, 1967–date.

PHYSICS

Applied Science and Technology Index. New York: Wilson, 1958–date.

POLITICAL SCIENCE

Social Science Index. New York: Wilson, 1974–date.

PAIS (electronic source).

PSYCHOLOGY
Psychological Abstracts. Lancaster, PA: APA, 1974–date.
PSYCLIT (electronic source).

RELIGION
Religion and Theological Abstracts. Chicago: ATLA, 1949–date.
Religion Index (electronic source).

SOCIOLOGY
Sociological Abstracts. New York: Sociological Abstracts, 1952–date.

WOMEN'S STUDIES
Women's Studies Abstracts. Rush, NY: Rush, 1972–date.

Specialized Dictionaries and Encyclopedias

Every discipline also has its own specialized dictionaries and encyclopedias. Such encyclopedias, especially, can provide useful articles. For example, if you were researching steroid use by athletes, you could check the *Encyclopedia of Sports.** If you were researching the work of American photographer Ansel Adams, you might consult the *Encyclopedia of American Art.*** Specialized dictionaries can also be handy to help translate difficult technical terms encountered as you read journal articles in the field. For a complete list of specialized dictionaries and encyclopedias in your subject area, consult the *Guide to Reference Books.*

Bibliographies

Another kind of specialized reference book available in most disciplines is the bibliographic index. Don't confuse this type of source with the bibliographic citations at the end of a book or article. That's a bibliography, too—a list of materials used by the author. A bibliographic index also lists books and articles but for a particular field of study, broken down according to various areas of interest. For example, the student working on the paper mentioned earlier about special problems faced by teenage children of alcoholics could look in the index at the back of the *Guide to Reference Books* under "Alcoholism" and find the page number for bibliographies on the subject. There's

**Encyclopedia of Sports* (New York: Barnes and Noble, 1978) (supplements).
***Encyclopedia of American Art* (New York: Dutton, 1981).

quite a list, including a bibliography of books and articles on alcohol and reproduction, another on alcohol education materials published in the United States, and many other lists of sources on areas of interest in the study of alcoholism. One bibliography seems particularly promising: *Alcoholism and Youth: A Comprehensive Bibliography,* by Grace Barnes.*

A good bibliography can save you a lot of work. Someone else has gone to the trouble of surveying the literature for you, and you can pick and choose what seems useful. One disadvantage is that often bibliographies do not include recent material. Obviously, check when the reference was published, and decide how far back you want to go.

Specialized Computer Databases

The bound reference book is quickly being replaced by the computer database, and in many ways, that's good. For instance, once you get past any lingering computerphobia, you'll find searching databases is easy and efficient. Many of the specialized indexes mentioned above are available both as bound books and on computer. There are also a number of additional specialized databases in each field to which your library may subscribe. For example, *ABI/INFORM* is a database in the business field that's available at a computer terminal in my campus library. *ERIC* is another database commonly available for researchers in education.

Ask your reference librarian what's available. A useful directory to computer databases, organized by subject or name, is:

Computer-Readable Databases: A Directory and Data Sourcebook. Edited by Martha E. Williams. Chicago: American Library Association, 1985–date (2 vols).

Internet Searches

If you haven't yet done a subject or keyword search of the World Wide Web, give it a try. Exercise 1.5, "A Quick Tour of the Internet," in Chapter 1, will show you how. If you have done a Web search, now consider using some of the search tools that will take you to parts of cyberspace the Web may not, including FTP space, WAIS databases, Gopher, and Telnet.

If you're accustomed to the ease of navigating the Web using its many graphic interfaces, you may find some of these search tools a bit

*Grace Barnes, *Alcoholism and Youth: A Comprehensive Bibliography* (Westport, CT: Greenwood, 1982).

Site	Search Tool	Web-Based Server
Gopher space	Veronica	gopher://veronica.scs.unr.edu/11//veronica
Telnet	Hytelnet	http://www.nova.edu/Inter-Links/cgi-bin/hytelnet.pl
FTP space	Archie	http://web.nexor.co.uk/archie.html
Usenet	DejaNews	http://www.dejanews.com
WAIS	WAISgate	http://www.wais.com

FIGURE 3.5 To increase your reach into cyberspace, use the appropriate search tool. Each can be accessed through the Web at the address listed above.

challenging at first. Most use text instead of graphics, and some, like Gopher, use hierachically organized directories instead of hypertext links. Most of these search tools, however, are accessible through the World Wide Web, either on all-in-one search pages or through other Web sites, including those listed in Figure 3.5.

MAPPING YOUR SEARCH: THE SECOND LEVEL

Document your dig into second-level sources here. Check the types of sources you've consulted, and list the specific titles.

☐ Specialized encyclopedias and dictionaries ____________

____

☐ Specialized journal indexes ____________

____

☐ Bibliographies ____________

____

☐ Computer databases ____________

____

☐ Internet search tools ____________

____

Third-Level Searching

By now, you're already way past the superficial searching for information you used to do for high school research papers. The *Reader's Guide* may already seem almost elementary. By the time you've completed first- and second-level searches for information, you've been pretty thorough. You've worked through general indexes as well as specialized indexes and encyclopedias and databases. If you're still not comfortable with how much information you've collected, move on to a third-level search or try expanding the search (see later in this chapter).

You may feel as if you have plenty of information to work with at this point. If so, it may be time to think about writing your paper. However, as you begin your draft next week or revise it later, you may discover that you had less material than you thought and need to do some more digging. A third-level search may be the place to start.

Search by Author

Typically, most searching is done by subject, but you may suddenly open up new doors if you begin using particular authors' names as keywords. Look for authors' names that repeatedly appear in bibliographies from articles and books you've found on your topic. These people are likely important authorities in the field and may have much more to say about your topic. Also note the names of authors whose work you particularly like or who seem influential. Use the card catalog and periodical indexes to see what else these authors have written. Is any of this information relevant?

Using Citation Indexes. Three citation indexes may also prove valuable when you have some authors' names to work with:

Arts and Humanities Citation Index. Philadelphia: Institute for Scientific Information, 1977–date.
Science Citation Index. Philadelphia: Institute for Scientific Information, 1961–date.
Social Science Citation Index. Philadelphia: Institute for Scientific Information, 1966–date.

Each of these indexes has three separate but related volumes: the *Source Index,* the *Citation Index,* and the *Permuterm Subject Index*.

Among other things, these indexes identify when the authors were cited *by other writers* in their own work, which may be relevant

```
HOBSON ES
  78 CONTRASTS BEHAVIOR A      219
     CALDWELL JP  COPEIA                  938  89
HOBSON EW
  14 J NAPIER INVENTION L
     BRYDEN DJ    ANN SCI          47    445  90
HOBSON GN
  69 J EXP PSYCHOL   80   386
     HEDLUND MA   J ANXIETY D       4    221  90
HOBSON J
  02 IMPERIALISM STUDY
     SHAPIRO H    WORLD DEV        18    861  90
  52 BRIT J ADDICT   49    5
     BERRIDGE V   BR J ADDICT      85    983  90  R
  77 AM J PSYCHIAT   134   97
     MOFFIT A     PSYCHIAT J       15     66  90
HOBSON JA
  ** DREAMING BRAIN
     TURNEY J     NEW STATE S       3     35  90  B
  1894 EVOLUTION MODERN CAP
  1894 MODERN CAPITALISM ST
     HOVENKAM H   STANF LAW R      42    993  90
  00 EC DISTRIBUTION
     ( ANON )     J ECON STUD      17     18  90
  00 WAR S AFRICA        233
     PORTER A     J AFR HIST       31     43  90
  01 SOCIAL PROBLEM           87
     FREEDEN M    ETHICS          100    489  90
  02 IMPERIALISM STUDY
     DUSSEL E     LAT AM PERS      17     62  90
  09 CRISIS LIBERALISM
     BEETHAM D    ARCH EUR SO      30    311  89  N
  11 SCI WEALTH
     HAINES WW    INT J SOC E      17     17  90
  12 CHARACTER LIFE      94
     FREEDEN M    ETHICS          100    489  90
  26 EVOLUTION MODERN CAP
     INIKORI JE   SOC SCI HIS      13    343  89
  29 EC ETHICS STUDY SOCI
     HOVENKAM H   STANF LAW R      42    993  90
  30 RATIONALISATION UNEM
     HOWARD MC    HIST POLIT       22     81  90
  31 LT HOBHOUSE HIS LIFE
     HELMESHA R   CAN R SOC A      27    357  90
  56 PHYSL IND
     SALTER J     HIST POLIT       22     65  90
  63 ANN INTERN MED   58   324
     SEE SCI FOR 1 ADDITIONAL CITATION
     DELVA NJ     BR J PSYCHI     157    703  90
     GLUSAC E     CAN J PSY        35    268  90  N
  65 EVOLUTION MODERN CAP
     KUHN R       AUST ECON H      26     53  88
  65 J PSYCHIAT RES           79
     ARANTES JM   ARQ BRAS P       41     77  89
  71 IMPERIALISM
```

Author

*Reference year (*means undated); title of Hobson's book*

J. Turney cites Hobson's book in review in journal New State S, *vol. 3, p. 35, 1990. Code letter at end for type of source (e.g., B—Book reviews, M—Meeting abstracts); if no code, source is article or report*

FIGURE 3.6 Entry in the *Social Science Citation Index* for author J. A. Hobson, listing works that cited him in 1990

Source: From Social Science Citation Index®. Copied with the permission of the Institute for Scientific Information®, 1992.

to your work. For example, when researching the purpose of dreaming, Christine came across an influential theory and book, *The Dreaming Brain,* by J. Allan Hobson,* that she wanted to discuss in her paper. The *Social Science Citation Index* helped her find other authors who mentioned Hobson in their articles, including a review of *The Dreaming Brain* (see Figure 3.6). By looking under "Hobson" in the *Source Index,* Christine was also able to find articles Hobson had written in a given year and a complete bibliography for each (see Figure 3.7).

The *Permuterm Subject Index* is the place to check if you don't have a specific author in mind. It will refer you to authors in the *Source Index*. The citation indexes take a little getting used to, but they're often full of helpful leads.

*J. Allan Hobson, *The Dreaming Brain* (New York: Basic Books, 1988).

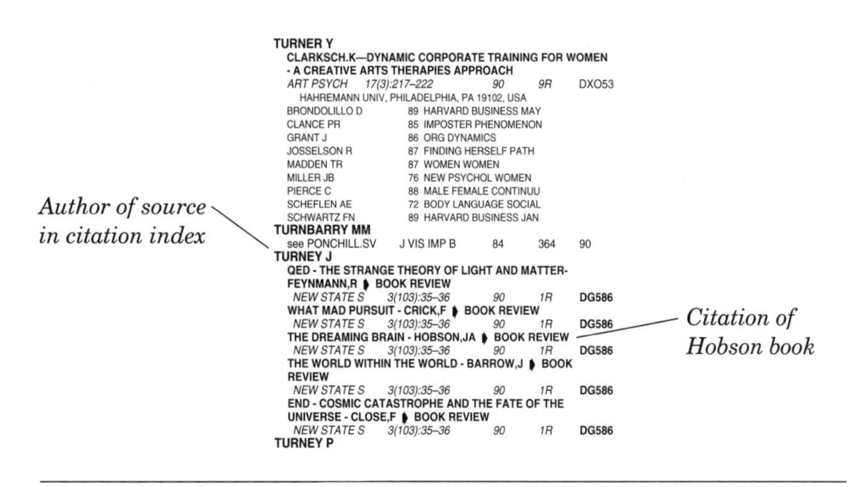

TURNER Y
CLARKSCH.K—DYNAMIC CORPORATE TRAINING FOR WOMEN
- A CREATIVE ARTS THERAPIES APPROACH
ART PSYCH 17(3):217–222 90 9R DXO53
HAHREMANN UNIV, PHILADELPHIA, PA 19102, USA
BRONDOLILLO D 89 HARVARD BUSINESS MAY
CLANCE PR 85 IMPOSTER PHENOMENON
GRANT J 86 ORG DYNAMICS
JOSSELSON R 87 FINDING HERSELF PATH
MADDEN TR 87 WOMEN WOMEN
MILLER JB 76 NEW PSYCHOL WOMEN
PIERCE C 88 MALE FEMALE CONTINUU
SCHEFLEN AE 72 BODY LANGUAGE SOCIAL
SCHWARTZ FN 89 HARVARD BUSINESS JAN
TURNBARRY MM
see PONCHILL.SV J VIS IMP B 84 364 90
TURNEY J
QED - THE STRANGE THEORY OF LIGHT AND MATTER-
FEYNMANN,R ▸ BOOK REVIEW
NEW STATE S 3(103):35–36 90 1R DG586
WHAT MAD PURSUIT - CRICK,F ▸ BOOK REVIEW
NEW STATE S 3(103):35–36 90 1R DG586
THE DREAMING BRAIN - HOBSON,JA ▸ BOOK REVIEW
NEW STATE S 3(103):35–36 90 1R DG586
THE WORLD WITHIN THE WORLD - BARROW,J ▸ BOOK
REVIEW
NEW STATE S 3(103):35–36 90 1R DG586
END - COSMIC CATASTROPHE AND THE FATE OF THE
UNIVERSE - CLOSE,F ▸ BOOK REVIEW
NEW STATE S 3(103):35–36 90 1R DG586
TURNEY P

FIGURE 3.7 The *Source Index* will provide a fuller description of a source mentioned in the citation index, including the bibliography (if any) of the article. (Turney's book review had no bibliography.)

Source: From Social Science Citation Index®. Copied with the permission of the Institute for Scientific Information®, 1992.

Bibliographies in Books and Articles

By now, you know that a bibliography in a published book or article can be a great help in your hunt for more information. A quick scan of the bibliography, even in a book that isn't exactly on your topic, can often suggest promising titles that are. Some books and articles have bibliographies, and others don't. Fortunately, there's a quick way to find out:

Bibliographic Index: A Cumulative Bibliography of Bibliographies. New York: Wilson, 1938–date.

This index, published annually, is organized by subject and lists books and journals published on that subject that feature bibliographies, including the pages where they can be found.

Unpublished Scholarly Papers

A vast amount of scholarly research is generated each year by graduate students working on advanced degrees. Much of this material is unpublished and therefore uncataloged in the usual indexes. But alas, there's an index to these sources, too:

Dissertation Abstracts International. Ann Arbor, MI: University Microfilms, 1970–date (from 1938–51, titled *Microfilm Abstracts,* and from 1952–69, *Dissertation Abstracts*).

This index, also available electronically, is worth checking. When I was writing my book on lobsters, I found a wonderful doctoral dissertation on a Maine island lobstering community that proved fascinating and useful.

First look for the subject/author index in a given year, which will then direct you to the *abstract,* or summary, of the paper in which you're interested. The abstract may be enough. If you want the entire paper, you can order it by mail through University Microfilms, Inc., Ann Arbor, MI 48106. (Copies, however, are costly—about $36.)

Essays and Articles Buried in Books

Somewhere, there may be an article or essay that's perfect for your paper, but you'll never find it because it's buried in a book about a related subject or the book's title doesn't sound promising when you come across it in the card catalog. How do you dig out this hidden gem? Check the following index:

Essay and General Literature Index. New York: Wilson, 1900–33, supplements 1934–date.

This index, also available on CD-ROM, covers the humanities and the social sciences and is easy to use. Search by subject or author. See the example in Figure 3.8.

MAPPING YOUR SEARCH: THE THIRD LEVEL

After you complete a third-level search, check the types of sources you consulted, and list the specific titles.

☐ Citation indexes ______________________________

____

☐ Dissertation indexes ______________________________

____

☐ Bibliographic indexes ______________________________

____

☐ *Essay and General Literature Index* ______________________________

____

Hurston, Zora Neale, 1907–1960

About

Bell, B. W. The Harlem Renaissance and the search for new modes of narrative. (*In* Bell, B. W. The Afro-American novel and its traditions p93-149)

Cooke, M. G. Solitude: the beginnings of self-realization in Zora Neale Hurston, Richard Wright, and Ralph Ellison. (*In* Cooke, M. G. Afro-American literature in the twentieth century p71-109)

Fox-Genovese, E. My statue, my self: autobiographical writings of Afro-American women. (*In* The private self; ed. by S. Benstock p63-89)

Johnson, B. Thresholds of difference: structures of address in Zora Neale Hurston. (*In* Johnson, B. A world of difference p172-83)

Neal, L. Eatonville's Zora Neale Hurston: a profile. (*In* Neal, L. Visions of a liberated future p81-96)

Pryse, M. Introduction: Zora Neale Hurston, Alice Walker, and the "ancient power" of black women. (*In* Conjuring; ed. by M. Pryse and H. J. Spillers p1-24)

About individual works

Their eyes were watching god

Awkward, M. "The inaudible voice of it all": silence, voice, and action in Their eyes were watching God. (*In* Awkward, M. Inspiriting influences p15-56)

Callahan, J. F. "Mah tongue is in mah friend's mouf": the rhetoric of intimacy and immensity in Their eyes were watching God. (*In* Callahan, J. F. In the African-American grain p115-49)

Dixon, M. Keep me from sinking down: Zora Neale Hurston, Alice Walker, and Gayl Jones. (*In* Dixon, M. Ride out the wilderness p83-120)

Gates, H. L. Color me Zora: Alice Walker's (re)writing of the speakerly text. (*In* Gates, H. L. The Signifying Monkey p239-58)

Gates, H. L. Color me Zora: Alice Walker's (re)writing of the speakerly text. (*In* Intertextuality and contemporary American fiction; ed. by P. O'Donnell and R. C. Davis p144-67)

Gates, H. L. Zora Neale Hurston and the speakerly text. (*In* Gates, H. L. The Signifying Monkey p170-216)

Johnson, B. Metaphor, metonymy, and voice in: Their eyes were watching God. (*In* Johnson, B. A world of difference p155-71)

Meese, E. A. Orality and textuality in Zora Neale Hurston's Their eyes were watching God. (*In* Meese, E. A. Crossing the double-cross p39-53)

Spillers, H. J. A hateful passion, a lost love. (*In* Feminist issues in literary scholarship; ed. by S. Benstock p181-207)

Wainwright, M. K. The aesthetics of community: the insular black community as theme and focus in Hurston's Their eyes were watching God. (*In* Harlem Renaisance: revaluations; ed. by A. Singh, W. S. Shiver, and S. Brodwin p233-43)

FIGURE 3.8 A student researching African-American author Zora Neale Hurston in the *Essay and General Literature Index* would find a listing of essays and articles about her and her work. Information such as this is often hard to find because the essays and articles are buried in collections.

Source: From *Essay and General Literature Index,* 1985–1989 Cumulative, Volume 11, pages 776–777. Copyright © 1989 by The H. W. Wilson Company. Material reproduced with permission of the publisher.

Expanding the Site of the Search

Library Sources

Sometimes, you're digging in the right place, but the site needs to be expanded. Students rarely take advantage of the enormous resources of their campus libraries. Most stick to the card catalog or the periodical indexes. But libraries contain some unusual sources of information, such as special collections and media departments, that can produce useful material on your topic.

Special Collections. Many university libraries are homes to unusual collections of material—historical documents, personal papers of prominent people, and the like—and may even be designated archives for significant works. Often, this material is catalogued separately. If your library has a special collection, consult the staff there about whether it may contain any useful material on your topic.

Audiovisual Departments. Audiovisual materials (films, records, tapes, slides, videocasettes, etc.) are available in separate departments in many libraries and are often catalogued separately. You may find some surprisingly useful sources on your topic in this catalog, such as a taped lecture by an authority who spoke on your campus or a film that offers some useful case studies.

Pamphlets. Pamphlet collections vary widely from library to library. Large libraries sometimes have extensive collections; small libraries, very few. If you have the time, you can send to various agencies for pamphlets relevant to your topic. Use the following index to find out what's available and who can provide it:

Vertical File Index: A Subject and Title Index to Selected Pamphlet Material. New York: Wilson, 1932/35–date.

Other Libraries. Your hometown library will likely not have a good collection of scholarly journals and books, but it may contain some useful sources that were perhaps missing or checked out from your university library's collection. Other public libraries in the area might be worth checking, too. Also remember that you can probably make use of interlibrary loan to get materials from other libraries without running all over the place to collect them yourself, though you'll need some lead time. (See Chapter 1, "Interlibrary Loan.")

Libraries and library resources across the United States and around the world are also accessible through the Internet. Hundreds of library catalogs, for example, are available using the Telnet protocol, but an increasing number of library resources can be reached through the Web, too. One site (and there are many) that offers convenient links to Web-based library materials is InterLinks. The InterLinks Library Resources page is located at the following address:

http://www.nova.edu/Inter-Links/library.html

Also consider other specialized libraries. Most museums have their own collections, which they often make available to college researchers. Historical societies and many large corporations have libraries, as well.

Nonlibrary Sources

It's so easy to forget that research doesn't begin and end at the library's front door. Be creative about other places to look!

Bookstores. Your campus library may not have recently published books on your topic (titles published in the last year), but your campus or local bookstore might. First, to find out what books seem promising that may not be listed in your library's card catalog, check one or more of the bibliographies of these trade books:

Books in Print. New York: Bowker, 1948–date. *Lists books by title and author.*
Paperbound Books in Print. New York: Bowker, 1955–date. *Lists only paperbacks.*
Subject Guide to Books in Print. New York: Bowker, 1957–date. *As the title implies, organized by subject.*

If you find a promising title, check local bookstores to find out if the book is in stock. If a book is not available and you have a week or ten days, you can also ask to order the book. Most bookstores will do this for you. The obvious drawback to this approach is that you have to buy the book you want. Doing so may very well be worth it, however, for a book that seems especially useful.

Writing Letters. There are people and organizations out there with exactly what you need. If you have enough lead time, you can write or call and have them send along some information. How do you find out who these people and organizations are, and how you can contact them?

First, ask people you know for information: your instructor, your friends, the people you interview for your paper. Also pay attention to the names of groups that crop up in your reading as well as the names of experts and the institutions they're associated with. Several reference sources can also help:

Encyclopedia of Associations. 32nd ed. Edited by Sandra Jaszczak. Detroit, MI: Gale Research, 1997. *Three parts; the last is a name and keyword index.*

Research Centers Directory. 22nd ed. Edited by Anthony Gerring. Detroit, MI: Gale Research, 1997. *Two volumes; covers 12,000 nonprofit and university-related research organizations.*

Both books will give you addresses and phone numbers of a variety of institutions and groups and, in many cases, names and phone numbers of key contact people (see Figure 3.9).

E-mail on the Internet also opens up new possibilities for reaching people and organizations. A growing number of groups have their own Web pages, which feature links to the e-mail addresses of infor-

★11863★ NARANON **(Substance Abuse)**
World Service Office
PO Box 2562
Palo Verdes, CA 90274 Phone: (213) 547–5800
Nonmembership. Provides assistance to drug-dependent individuals and their families. Organizes discussion groups. **Telecommunications Services:** Telephone referral service. Group is distinct from Narcotics Anonymous (see separate entry) located in Van Nuys, CA.

★11863★ NARCOTIC EDUCATIONAL FOUNDATION OF AMERICA **(Substance Abuse)** (NEFA)
5055 Sunset Blvd. Phone: (213) 663–5171
Los Angeles, CA 90027 Henry B. Hall, Exec. Dir.
Founded: 1924. To provide education about narcotics and other drugs in order to warn youth and adults about the dangers of drug abuse. Has produced films and maintains film library. Operates speakers' bureau, counseling and referral service, library and reading room. Conducts research. Helps produce television and radio programs.

Publications: *Get the Answers - An Open Letter to Youth, Some Things You Should Know About Prescription Drugs, Drugs and the Automotive Age,* and other *Student Reference Sheets* outlining the dangers associated with the use of various substances.

FIGURE 3.9 The *Encyclopedia of Associations* will provide a researcher writing about drug abuse treatment, for example, the names of several organizations to write to or call for information.

Source: Taken from *Encyclopedia of Associations,* 26th edition, Volume 1. Edited by Deborah M. Burek. Copyright © 1991 by Gale Research. Reproduced by permission.

mation officers, authors of online documents, and other informed individuals. You can also sometimes unearth the e-mail addresses of experts on your topic using search tools designed for that purpose. For tips on how to do this, check out "Finding People on the Internet" in Chapter 2.

Lectures. Every week on my campus, there are ten or more public lectures on a variety of subjects, ranging from the biodynamics of the rain forest to the dangers of date rape. A lecture on your topic—or a closely related one—could be a boon to your research, providing not only fresh material that often has a local angle but also a live person to quote and interview. Going to lectures on the hope of finding useful information is a hit-or-miss approach. Nonetheless, keep your eye on the listings of public lectures in your campus newspaper.

TV and Radio. Most people in the United States get most of their information about public issues from television. That doesn't mean TV is the best source of information, but it is certainly an influential one. Television and radio news, public affairs programs, and even talk shows can be useful sources of information. *TV Guide* can be a useful reference, and so can the local newspaper, which may list the topics discussed on various television talk shows that day.

MAPPING YOUR SEARCH: EXPANDING THE SEARCH

Document the ground you covered when you expanded your search here.

☐ Special collections

☐ Audiovisual department

☐ Pamphlets

☐ Other libraries ______________________________

____

____

☐ Bookstores

☐ Lectures

☐ TV and radio

Conducting Interviews

You've already thought about whether interviews might contribute to your paper. Last week, you began to build a list of possible interview subjects and probably contacted several of them. By the end of this week, you should begin interviewing.

I know. You wouldn't mind putting it off. But once you start, it will get easier and easier. I should know. I used to dread interviewing strangers, but after making the first phone call, I got some momentum going, and I began to enjoy it. It's decidedly easier to interview friends, family, and acquaintances, but that's the wrong reason to limit yourself to people you know.

Whom to Interview? Interview people who can provide you with what you want to know. And that may have changed after your research this week. In your reading, you might have encountered the names of experts you'd like to contact, or you may have decided that what you really need is some anecdotal material from someone with experience in your topic. It's still not too late to contact interview subjects who didn't occur to you last week. But do so immediately.

What Questions to Ask? The first step in preparing for an interview is to ask yourself, What's the purpose of this interview? In your research notebook, make a list of *specific questions* for each person you're going to interview. Often, these questions are raised by your reading or other interviews. What theories or ideas encountered in your reading would you like to ask your subject about? What specific facts have you been unable to uncover that your interview subject may provide? What don't you understand that he could explain? Would you like to test one of your own impressions or ideas on your subject? What about the subject's work or experience would you like to learn? Interviews are wonderful tools for clearing up your own confusion and getting specific information that is unavailable anywhere else.

Now make a list of more *open-ended questions* you might ask each or all the people you're going to talk to. Frankly, these questions are a lot more fun to ask because you're more likely to be surprised by the answers. For example:

- In all your experience with ________________, what has most surprised you?
- What has been the most difficult aspect of your work?
- If you had the chance to change something about how you approached ________________, what would it be?

- Can you remember a significant moment in your work on _____________? Is there an experience with _____________ that stands out in your mind?
- What do you think is the most common misconception about _______________? Why?
- What are significant current trends in _______________?
- Who or what has most influenced you? Who are your heroes?
- If you had to summarize the most important thing you've learned about _______________, what would it be? What is the most important thing other people should know or understand?

As you develop both specific and open-ended questions, keep in mind what you know about each person—his work in the field and personal experience with your topic. You may end up asking a lot of the same questions of everybody you interview, but try to familiarize yourself with any special qualifications a subject may have or experiences he may have had. That knowledge might come from your reading, from what other people tell you about your subject, or from your initial telephone call to set up the interview.

Also keep in mind the *kinds* of information an interview can provide better than other sources: anecdotes, strong quotes, and sometimes descriptive material. If you ask the right questions, a live subject can paint a picture of his experience with your topic, and you can capture that picture in your paper.

During the Interview. Once you've built a list of questions, be prepared to ignore it. Interviews are conversations, not surveys. They are about human interaction between two people who are both interested in the same thing.

I remember interviewing a lobsterman, Edward Heaphy, on his boat. I had a long list of questions in my notebook, which I dutifully asked, one after the other. My questions were mechanical, and so were his answers. I finally stopped, put my notebook down, and talked informally with Edward for a few minutes. Offhandedly, I asked, "Would you want your sons or daughter to get in the business?" It was a totally unplanned question. Edward was silent for a moment, staring at his hands. I knew he was about to say something important because, for the first time, I was attentive to him, not my notepad. "Too much work for what they get out of it," he said quietly. It was a surprising remark after hearing for the last hour how much Edward loved lobstering. What's more, I felt I had broken through. The rest of the interview went much better.

Much of how to conduct an interview is common sense. At the outset, clarify the nature of your project—what your paper is on and where you're at with it. Briefly explain again why you thought this individual would be the perfect person to talk to about it. I find it often helps to begin with a specific question that I'm pretty sure my subject can help with. But there's no formula. Simply be a good conversationalist: listen attentively, ask questions that your subject seems to find provocative, and enjoy with your subject sharing an interest in the same thing. Also don't be afraid to ask what you fear are obvious questions. Demonstrate to the subject that you *really* want to understand.

Always end an interview by making sure you have accurate background information on your subject: name (spelled correctly), position, affiliation, age (if applicable), phone number. Ask if you can call him with follow-up questions, should you have any. And always ask your subject if he can recommend any additional reading or other people you should talk to. Of course, mention that you're appreciative of the time he has spent with you.

Notetaking. There are basically three ways to take notes during an interview: use a tape recorder, a notepad, or both. I adhere to the third method, but it's a very individual choice. I like tape recorders because I don't panic during an interview that I'm losing information or quoting inaccurately, but I don't want to spend hours transcribing the tapes. So I also take notes on the information I think I want to use, and if I miss anything, I consult the recording later. It's a backup. Sometimes, I find that there is no recording—the machine decided not to participate in the interview—and at least I have my notes. Again, a backup.

Get some practice developing your own notetaking technique by interviewing your roommate or taking notes on the television news. Devise ways to shorten often-used words (e.g., *t* for *the, imp* for *important,* and *w/o* for *without*).

Conducting Surveys

Last week, you considered whether your topic lends itself to an informal survey. If so, you generated three types of questions you might ask: *open ended, multiple choice,* and *directed* (see "Survey Design" in Chapter 2). After all the reading you did this week, you likely have some fresh ideas of questions you might ask. Finalize the questions, and begin distributing the survey to the target group you defined earlier.

Distribution. Surveys administered by telephone have some advantages. People are more likely to be direct and honest over the phone, since they are relatively anonymous. Surveys are also more likely to be completed correctly, since the answers are recorded by the survey giver. However, making multiple phone calls can be tedious and expensive, if your target group goes beyond the toll-free calling area. But you may have no choice, especially if the target group for your survey isn't exclusively on campus.

The alternative to conducting a telephone survey is to distribute the survey yourself. The university community, where large numbers of people are available in a confined area, lends itself to administering surveys this way, if there's a university audience you're interested in polling. A survey can be distributed in dormitories, dining halls, classes, or anywhere else the people you want to talk to gather. You can stand outside the student union and stop people as they come and go, or you can hand out your survey to groups of people and collect them when the participants have finished. Your instructor may be able to help distribute your survey to classes. I asked a number of my colleagues to distribute Christine's survey (see Figure 2.3) in their Freshman English classes, a required course representing a relatively random sample of freshmen. Since the survey only took five minutes to fill out, other instructors were glad to help, and in one day, Christine was able to sample more than ninety students.

The campus and its activities often self-select the group you want to survey. Anna, writing a paper on date rape, surveyed exclusively women on campus, many of whom she found in women's dormitories. For his paper on the future of the fraternity system, David surveyed local "Greeks" at their annual awards banquet.

How large a sample should you shoot for? Since yours won't be a scientific survey, don't bother worrying about statistical reliability; just try to survey as many people as you can. Certainly, a large (say, more than one hundred) and representative sample will lend more credence to your claims about any patterns observed in the results.

LOOKING BACK BEFORE MOVING ON

You've just completed a crucial week in your research project. If you've been thorough in your hunt for information and thoughtful about taking notes on what you find, your paper is more than half completed. But remember, the research process—like the writing process—is *recursive.* That is, you may find yourself circling back to previous steps, especially those in the section "Digging Deeper for Information."

Before you move on, it may help to quickly trace where you've been searching and what you've found this week. Any steps you've missed or had trouble with may be the place to start next week. You may also want to show your instructor the trail you've followed. He might suggest some other places to look.

4

□ ■ □

The Fourth Week

GETTING TO THE DRAFT

It is *not* 2 A.M. Your paper is *not* due in twelve hours but in one or two weeks. For some students, beginning to write a research paper this early—weeks before it's due—will be a totally new experience. An early start may also, for the first time, make the experience a positive one. I know that starting early will help ensure writing a better paper.

Still, there are those students who say they thrive on a looming deadline, who love working in its shadow, flirting with failure. "I work best that way," they say, and they wait until the last minute and race to the deadline in a burst of writing, often sustained by cigarettes and strong doses of caffeine. It works for some students. Panic is a pretty strong motivator. But I think most who defend this habit confuse their relief at successfully pulling off the assignment once again with a belief that the paper itself is successful.

Papers done under such pressure often aren't successful, and that is particularly true of the last-minute research paper, where procrastination is especially deadly. Research writing is recursive. You often have to circle back to where you've already been, discovering holes in your research or looking at your subject from new angles. It's hard to fit in a trip back to the library the night before the paper is due, when you've just started the draft and need to check some information. This book is designed to defeat procrastination, and if, in the past few weeks, you've done the exercises, taken thoughtful notes, and attempted a thorough search for information, you probably have the urge to begin writing.

On the other hand, you may feel as if you don't know enough yet about your topic to have anything to say. Or you may be swamped with information, and your head may be spinning. What do you do with it all?

When Christy came to my office, she was three weeks into her research on a paper that asked, Why do diets fail? She really wanted to know, since she was having such a hard time with her own diet. Though she'd really done a good job collecting information, she was exasperated.

"I found a whole bunch of articles on how heredity affects obesity," she said, "and all this stuff on how people's upbringing determines how they eat. I also found some articles that said our bodies *want* to be certain weights."

It sounded pretty interesting to me.

"I've got all this information, but I'm worried that I'll lose my focus," she said. "*And so much of it seems contradictory*. I don't know what to think."

When the Experts Disagree

Christy was pretty sure she was in trouble because her sources sometimes didn't agree on the same things. I thought she was right where she should be: standing on the curb at a busy intersection, watching the experts on her topic collide and then go off in different directions. Knowledge in any field—nutrition, literature, or entomology—is not static. It is contested—pushed, pulled, probed, and even sometimes turned over completely to see what is underneath. Scholars and experts devote their lifetimes to disagreeing with each other, not because they enjoy being disagreeable but because when knowledge is contested, it is advanced.

When I researched lobsters, I discovered a fascinating scientific mystery: More than 90 percent of the lobsters that grow to the minimum legal size every year end up on someone's dinner table. At that size, most lobsters haven't even had a chance to breed. How is it possible, asked the scientists, that there are any lobsters left at that rate of exploitation? I discovered several explanations. Some people argued that the millions of lobster traps—each of which is designed to allow undersize lobsters to escape—serve as a kind of giant soup kitchen, providing extra food to lobsters. That, some experts said, accounted for lobster's resilience. Other experts believed that laws protecting females carrying eggs have worked remarkably well. Still others believed that lobsters migrate into areas depleted by overfishing. Recently, another idea won favor with scientists. They suggested that large lobsters at the edge of the continental shelf are the "parental stock" for coastal lobsters, sending their larval offspring inshore on tides and currents.

Evaluating Conflicting Claims

As a writer—and in this case, a nonexpert—I had to sort through these conflicting opinions and decide which I thought were most convincing. I had to claim my point of view and later make it convincing to my own readers.

That was Christy's challenge, and it's your challenge, too. When you're thorough in your research, you're bound to find sources that square off against each other or come at your subject from different directions. What do you make of these competing claims and differing perspectives?

❑ *EXERCISE 4.1*

Pointed or Pointless: How Do You Decide What's True?

Is boxing dangerous and pointless, or is it risky but getting safer? Suppose you were confronted with the following conflicting claims about boxing. How would you judge the relative truth of each? On what basis would you be inclined to believe one and not the other? What else might you need to know to make that decision?

As you discuss these issues in class, pay attention to the differing nature of each claim and *how* it is made.

CLAIM	COUNTERCLAIM
▪ Boxing is the world's most dangerous—and pointless—sport.	▪ Boxing, like all sports, has risks, but advances in gear have made it safer than ever.
▪ Eighty-seven percent of boxers suffer from brain damage.	▪ Most boxers are amateurs and rarely fight long enough to get hurt.
▪ Eighty-seven percent of all professional boxers suffer from brain damage.	▪ Fewer than half of professional boxers suffer from minor neurological disruptions in the brain area.
▪ Muhammed Ali is a classic example of a fighter who took one too many punches.	▪ George Foreman is still fighting at middle age and is not only still agile in the ring but smart enough to make millions hawking Midas mufflers.

□ ■ □

When confronted with contradictory evidence, the easiest thing to do is simply suppress the views that you don't agree with. You can do that if the views lack credibility, but even incredible ideas can *become* credible. Some years ago, the idea that the continents were essentially floating plates, which through the millennia had drifted apart, was viewed with great skepticism. Now plate tectonics is a chapter in all standard geology texts. Other ideas—like the assertion by neo-Nazi groups that the Holocaust didn't happen—are clearly outrageous. But a discussion of those ideas sometimes can be instructive, as well.

Before you dismiss or suppress any idea that seems at odds with what you think, consider the source and the strength of the argument. Here are some suggestions that may help you evaluate both:

1. Who is making the claim? Is it someone who is a respected authority in the field?
2. How convincing is the evidence he provides to support the assertions?
3. Do other credible sources lend support to the claim?
4. What do your own experiences and observations (if any) suggest about what a source is saying? What makes sense to you?
5. How many people believe the assertion? Even if a claim seems wrong, is it influential?
6. What does the advocacy of a clearly wacky idea say about the nature of the problem you're researching?

❑ *EXERCISE 4.2*
Reclaiming Your Topic

More than two weeks ago, you began researching a topic that you may have known little about. But you were curious enough to dive in and immerse yourself in the research, listening to the voices of people who know more than you. You may feel, as Christy did, that your paper is beginning to slip away from you; there is just too much information, or the contradictions can't possibly be sorted out. It might seem presumptuous to think that your ideas matter. You may feel as if you're in over your head. After all, you're not an expert.

If you're not at all confused at this stage in the research process, that's great. Now is the time, through writing, to tighten your grasp on the material. But if you're feeling overwhelmed, writing now can help you get a grip. Try this exercise, which will take about forty minutes.

Step 1: Spend ten or fifteen minutes reviewing all of the notes you've taken so far and skimming key articles or passages from books. Glance at your most important sources. Let your head swim with information.

Step 2: Now clear your desk of everything but your research notebook. Remove all your notes and all your sources. You won't use them while doing the rest of this exercise. Trust that you'll remember what's important.

Step 3: Now fastwrite about your topic for eight minutes. Tell the story of how your own thinking about your topic has evolved. When you began the project, what did you think? Then what happened, and what happened after that? What were your preconceptions about your topic? How have they changed? This is an open-ended fastwrite. Don't let the writing stall out. If you run out of things to say, talk to yourself through writing about your research, thinking about other trails you might follow. Time yourself.

Step 4: Skip a few lines in your notebook. Write "Moments, Stories, People, and Scenes." Now fastwrite for another ten minutes, this time, focusing on more specific case studies, situations, people, experiences, observations, and so on that stand out in your mind from the research done so far or perhaps from your own experience with the topic. Keep your pen moving for a full ten minutes. Time yourself.

Step 5: Skip a few more lines. For ten minutes, quickly write a dialogue between you and someone else about your topic. You choose whom to converse with—a friend, your instructor. Don't plan the dialogue. Just begin with the question most commonly asked about your topic, and take the conversation from there, writing both parts of the dialogue.

Step 6: Finally, skip a few more lines and write these two words in your notebook: "So What?" Now spend a few minutes trying to summarize the most important thing *you* think people should understand about your topic based on what you've learned so far. Distill these comments down to a sentence or two. This may be hard, but it's important. Remember, you can change your mind later.

What did doing this exercise accomplish, besides giving you a cramp in your writing hand? If the exercise worked, you probably already know. By freeing yourself from the chorus of expert voices in your sources and thinking to yourself about what the ideas you've collected mean, you've taken possession of the information again. You may have reaffirmed your purpose in writing the paper.

It may help you grasp the meaning of this exercise—and what completing it can do for you—by looking at how another student, Candy,* found her purpose in writing. After reviewing her notes and materials (Step 1) and then putting them away in preparation to write (Step 2), Candy was ready for Step 3, the first fastwrite. She told the story of finding her focus—how child abuse affects language development—by noting things that struck her as she went along:

STEP 3

Well, let's see, in the beginning, I was going to do the effects in general of child abuse. As I was researching this, I discovered that I would have to narrow it down because there was so much information on the general effects of child abuse. Initially, when I came across the idea that child abuse creates an impairment in speech and language development, I almost just threw it out. But, I went ahead and read the article. It was very interesting and I was able to relate to it. I have taken a course in linguistics so, I was able to relate to how this could be possible. So, I looked further into the topic of the effects that child abuse has on the language development of children and found quite a bit of information. It became more and more interesting to me as I read the information and all the tests that have been run to prove this idea. Before I began research, I never thought that this could be a possible effect of child abuse. But, after researching and thinking more and more about it, I find it quite logical.

By focusing on specifics in Step 4, the second fastwrite, you should discover some ways to anchor your ideas about the topic to particular people, situations, and case studies you discovered in your

*The following excerpts are reprinted with permission of Candyce C. Collins.

reading or from your own experience. Making these connections will not only strengthen your own thinking; case studies and personal accounts often make compelling examples, important to your paper.

In doing Step 4, Candy recalled the story of Genie, a girl who was confined to a closet by an abusive parent until she was thirteen. Genie later became a case used at the beginning and ending of Candy's paper. Here's Candy's second fastwrite:

STEP 4

One case study that stands out is a story about Genie. This little girl had an extreme case of child abuse and neglect when she was a child up until she was 13 when she was found. At 13, she spoke nothing. Now, this is a severe case of language deficiency as a result of child abuse but it goes on to show that it happens. She was locked away in her room, tied to her crib. Her father would beat her when ever she made a noise, so as she got older, she feared to say or utter a sound, so she didn't. As a result, she was unable to talk. She was never brought out in the world, never watched TV or heard the radio.

Other studies have been done on groups of abused, neglected, and both abused and neglected children. The tests prove that all three groups showed signs of slower language development when compared to nonabused children. The highest results were found in the neglected only children, then the both neglected and abused and lastly the abused only children. This is due to the fact that the two groups of children that were abused, had some stimulation in their lives even though it might not have been pleasant.

Step 5, the dialogue writing activity, invites someone else to the discussion of your topic, challenging you to consider an audience. What might most people want to know about your topic? How might you explain the answers? These questions may later shape how you organize your paper.

Candy's dialogue started with the question that began her research—What are some of the effects of child abuse?—and then went from there, getting more and more specific. Can you visualize the inverted pyramid progression of her questions and answers? Candy later used this form in part of her paper.

It actually might be more productive to construct a more freewheeling dialogue than Candy's. Have a real conversation with an imagined reader. Push yourself with questions that really get you thinking about your topic and that might help you see it in a fresh way.

Here's Candy's dialogue:

STEP 5

What are some of the effects that children suffer from as a result of child abuse?

Well, there's lots of them. One in particular that most don't think of is that child abuse can cause language development problems in children.

What kind of language problems do they have?

Primarily, they lack the ability to communicate as well with others than do the nonabused children. Studies show that they have a distinct style of communication. One that is more aggressive and hostile and they try to avoid any true contact through conversations. In extreme cases, like one girl named Genie whose dad beat her whenever she made a sound, couldn't speak at all by the age of 13.

What causes this development problem with language?

When it comes down to it, the main reason is that these children are lacking the normal stimulation that they should receive. They're not exposed to the experiences that would be necessary to learn new words. Also, they are deprived of the parent-child relationship that is an important part of the language acquisition process.

What types of abuse are you talking of? Is it all kinds?

No, it's not all kinds of abuse that cause this. It is limited to those children who were either neglected, physically abused or both. Primarily, those that were solely neglected suffer the most. It has been proven that sexual abuse doesn't have an adverse effect on language but in fact these children seem more mature as far as language goes. But, I didn't research into that.

Finally, asking "So What?" in Step 6 should help you redefine your thesis, or the controlling idea of your paper. In fact, your thesis may change. But for now, you need some brief statement—a sentence or two—that summarizes the most important thing you want your readers to understand.

Candy's answer to the "So What?" question later became the main point of her paper:

STEP 6

"So What?"

Child abuse has a negative effect on the language development of a child. This is a result of the child's lack of stimulation, interaction, experiences, and parent-child relationships, which are all essential to the proper development of language.

If you're not happy with your answer to "So What?" spend some more time thinking about it. Don't proceed too much further with writing until you have some kind of tentative thesis statement to keep in mind. Put your thesis on an index card or piece of paper, and post it over your desk as a reminder. Pull it down and revise it, if necessary, as you continue with research and writing. But keep that thesis up there on the wall, at least while you're writing the first draft.

□ ■ □

If Exercise 4.2 didn't work for you, you may need to collect more information. Consider circling back to some of the suggestions made in the third week in "Digging Deeper for Information" (see Chapter 3). But if you feel ready to begin writing a draft, read on.

Deciding Whether to Say *I*

I'm a writer who seems unable to stop talking about myself. As a reader of this textbook, that should be apparent to you by now. I share anecdotes about my photography failures, my high school girlfriend, and my predilection for lobsters. I've chosen to do this, though I know that getting personal in a piece of writing is somewhat risky business. If it's excessive, self-disclosure can seem egotistical or narcissistic. Constant self-reference—"I believe that . . ." or "I always wondered about . . ." or "I feel that . . ."—is usually unnecessary. (After all, if you simply make the assertion without the attribution, it's pretty obvious that you believe it or feel it.) The overuse of *I* can also seem to get in the way of the real subject, which may not be you. The personal profile is one genre of nonfiction writing that often suffers from explicit authorial intrusion. And teachers of research papers, as you know, often seem downright hostile to the intruding *I*.

By now, you know I don't agree with the view that all research writing should be objective (as if such a thing were possible). And in the research *essay* that you are about to draft this week, I certainly invite you to consider using the first person, presenting your own observations and experiences as evidence (if they're relevant) and yes, even talking about yourself.

There are many reasons this might be a good idea. First, by signaling our personal experiences and prejudices about a topic, we make explicit not only our particular purposes in exploring it but why we might have a reason for (or even a vested interest in) seeing it a certain way. Readers like to know a writer's motivation for writing about something and appreciate knowing how her experiences might influence her ways of seeing. But maybe even more important, when a writer stops pretending that the *text* talks instead of the *author* (e.g., "This paper will argue that . . .") and actually enters into her text, she is much more likely to initiate a genuine conversation with her readers *and* with her sources. This dialogue might very well lead to some new ways of seeing her topic—that is, after all, the purpose of inquiry.

Getting *Personal Without* Being *Personal*

Conversation takes place between people, and in writing that embodies conversation, readers sense what Gordon Harvey* called *presence*—an awareness that a writer is making sense of things in his own particular ways, that he has a personal stake in what is being said. This is most easily achieved when the writer *gets* personal by using the first person, sharing personal experiences and perspectives.

*Gordon Harvey, "Presence in the Essay," *College English* 56 (1994): 642–654.

I hope that you sense my presence in *The Curious Researcher* through my willingness to do such things.

But I also want you to see, as Harvey observes, that presence in writing can be registered in ways other than simply talking about yourself. That is, you can write a research essay this week that *doesn't* use the first person or isn't autobiographical and still provides your readers with a strong sense of your presence as an individual writer and thinker. This presence may be much more subtle when it's not carried on the first-person singular's sturdy back. But it still makes writing come to life.

❑ *EXERCISE 4.3*
Presence in the Research Essay*

Step 1: Find Carolyn Nelson's research essay on the Endangered Species Act in Appendix B. This is a strong essay for a number of reasons, one of which is that Carolyn registers her presence without ever using the first person or including autobiographical details. She certainly could have, but she decided that, in this case, it would not serve her purpose.

Step 2: Your instructor will either read all or part of Carolyn's essay aloud, or he may ask you to read it to yourself. As you listen or read, put an *A* (for *Author*) next to lines or passages in Carolyn's essay in which you sense her presence most strongly. Where do you sense she is expressing her personality, her feelings, her own particular ways of seeing things?

Step 3: Tally the responses of the entire class. Which paragraphs or passages garnered the most *A*'s? Which had the fewest?

Step 4: In small groups or as a full class, discuss *how* Carolyn manages to register her presence in her writing without talking about herself or using the first-person singular. Look at those passages which earned the most *A*'s: What are they like? Build a list of ways a writer can make herself felt by a reader without explicit self-disclosure. If you have time, also discuss those passages in which Carolyn seems to lack presence. What do you notice about the characteristics of those passages?

*This exercise was inspired by Professor Tony Nevin, a friend, colleague, science writer, and former student.

Before you begin drafting your essay this week, you'll have to decide how you'd prefer to get personal—explicitly or implicitly. For some of you, the choices may be limited. For instance, if your essay is on the causes of World War I, then integrating your own personal experience with the subject is obviously not an option. Most topics, however, offer the possibility of self-disclosure, and unless your instructor advises otherwise, almost all can accommodate *I*. But as I hope the preceding exercise demonstrated, when you choose not to get personal in direct ways, you can still establish a strong presence in your essay.

Beginning at the Beginning

John McPhee, a staff writer for *New Yorker* magazine and one of the masters of writing the research-based essay, gave a talk some years back about beginnings, which vex many writers.

> The first part—the lead, the beginning—is the hardest part of all to write. I've often heard writers say that if you have written your lead you have written 90 percent of the story. You have tens of thousands of words to choose from, after all, and only one can start the story, then one after that, and so forth. And your material, at this point, is all fresh and unused, so you don't have the advantage of being in the middle of things. You could start in any of many places. What will you choose?

Leads must be sound. They should never promise what does not follow. Leads, like titles, are flashlights that shine down into the story.*

Flashlights or Floodlights?

I love this: *"Leads . . . are flashlights that shine down into the story."* An introduction, at least the kind I was taught to write in high school, is more like a sodium vapor lamp that lights up the whole neighborhood. I remember writing introductions to research papers that sounded like this:

```
There are many critical problems that face
society today. One of these critical problems is
environmental protection, and especially the
```

*John McPhee, University of New Hampshire, 1977.

conservation of marine resources. This paper will explore one of these resources--the whale--and the myriad ways in which the whale-watching industry now poses a new threat to this species' survival. It will look at what is happening today and what some people concerned with the problem hope will happen tomorrow. It will argue that new regulations need to be put into effect to reduce boat traffic around our remaining whales, a national treasure that needs protection.

This introduction isn't that bad. It does offer a statement of purpose, and it explains the thesis. But the window it opens on the paper is so broad—listing everything the paper will try to do—that readers see a bland, general landscape. What's to discover? The old writing formula for structuring some papers—"Say what you're going to say, say it, and then say what you said"—breeds this kind of introduction. It also gets the writer started on a paper that often turns out as bland as the beginning.

Consider this alternative opening for the same paper:

Scott Mercer, owner of the whale-watching vessel Cetecea, tells the story of a man and his son who decide that watching the whales from inside their small motorboat isn't close enough. They want to swim with them. As Mercer and his passengers watch, the man sends his son overboard with snorkel and fins, and the boy promptly swims towards a "bubble cloud," a mass of air exhaled by a feeding humpback whale below the surface. What the swimmer didn't know was that, directly below that bubble cloud, the creature was on its way up, mouth gaping. They were both in for a surprise. "I got on the P.A. system and told my passengers, just loud enough for the guy in the boat to hear me, that either that swimmer was going to end up as whale food or he was going to get slapped with a $10,000 fine. He got out of the water pretty fast."

I think this lead accomplishes nearly as much as the bland version but in a more compelling way. It suggests the purpose of the paper—to explore conflicts between whale lovers and whales—and even implies the thesis—that human activity around whales needs more regulation. This lead is more like McPhee's "flashlight," pointing to the direction of the paper without attempting to illuminate the entire subject in a paragraph. An interesting beginning will also help launch the writer into a more interesting paper, for both reader and writer.

It's probably obvious that your opening is your first chance to capture your reader's attention. But how you begin your research paper will also have a subtle yet significant impact on the rest of it. The lead starts the paper going in a particular direction; it also establishes the *tone,* or writing voice, and the writer's relationships to the subject and the reader. Most writers at least intuitively know this, which is why beginnings are so hard to write.

Writing Multiple Leads

One thing that will make it easier to get started is to write three leads to your paper, instead of agonizing over one that must be perfect. Each different opening you write should point the "flashlight" in a different direction, suggesting different trails the draft might follow. After composing several leads, you can choose the one that you—and ultimately, your readers—find most promising.

Writing multiple openings to your paper might sound hard, but consider all the ways to begin:

- *Anecdote.* Think of a little story that nicely frames what your paper is about, as does the lead about the man and his son who almost became whale food.
- *Scene.* Begin by giving your readers a look at some revealing aspect of your topic. A paper on the destruction of tropical rain forests

might begin with a description of what the land looks like after loggers have left it.

▪ *Profile.* Try a lead that introduces someone who is important to your topic. Candy's lead, using a case study on Genie, the abused thirteen-year-old, is a good example. So is Christina's profile of her own struggle to emote on stage in an essay on method acting (see Appendix A).

▪ *Background.* Maybe you could begin by providing important and possibly surprising background information on your topic. A paper on steroid use might start by citing the explosive growth in use by high school athletes in the last ten years. A paper on a novel or an author might begin with a review of what critics have had to say.

▪ *Quotation.* Sometimes, you encounter a great quote that beautifully captures the question your paper will explore or the direction it will take. Heidi's paper on whether *Sesame Street* provides children with a good education began by quoting a tribute from *U.S. News and World Report* to Jim Henson after his sudden death.

▪ *Dialogue.* Open with dialogue between people involved in your topic. Dan's paper on the connection between spouse abuse and alcoholism began with a conversation between himself and a woman who had been abused by her husband.

▪ *Question.* Pointedly ask your readers the questions you asked that launched your research or the questions your readers might raise about your topic. Here's how Kim began her paper on adoption: Could you imagine going through life not knowing your true identity?

▪ *Contrast.* Try a lead that compares two apparently unlike things that highlight the problem or dilemma the paper will explore. Dusty's paper "Myth of the Superwoman" began with a comparison between her friend Susan, who married at 21 and grew up believing in Snow White and Cinderella, and herself, who never believed in princes or white horses and was advised by her mother that it was risky to depend on a man.

▪ *Announcement.* Sometimes the most appropriate beginning *is* one like the first lead on whales and whale-watchers mentioned earlier, which announces what the paper is about. Though such openings are sometimes not particularly compelling, they are direct. A paper with a complex topic or focus may be well served by simply stating in the beginning the main idea you'll explore and what plan you'll follow.

❑ *EXERCISE 4.4*
Three Ways In

Step 1: Compose three different beginnings, or leads, to your research paper. Each should be one or two paragraphs (or perhaps more, depending on what type of lead you've chosen and on the length of your paper). Think about the many different ways to begin, as mentioned earlier, and experiment. Your instructor may ask you to write the three leads in your research notebook or type them on a separate piece of paper and bring them to class.

Step 2: Get some help deciding which opening is strongest. Circulate your leads in class, or show them to friends. Ask each person to check the one lead he likes best, that most makes him want to read on.

Step 3: Choose the lead you like (even if no one else does). To determine how well it prepares your readers for what follows, ask a friend or classmate to answer these questions: Based on reading only the opening of the paper: (a) What do you predict this paper is about? What might be its focus? (b) Can you guess what central question I'm trying to answer? (c) Can you predict what my thesis might be? (d) How would you characterize the tone of the paper?

It's easy to choose an opening that's catchy. But the beginning of your paper must also help establish your purpose in writing it, frame your focus, and perhaps even suggest your main point, or thesis. The lead will also establish the voice, or tone, the paper will adopt (see the following section). That's a big order for one or two paragraphs, and you may find that more than a couple of paragraphs are needed to do it. Tentatively select the one opening (or a combination of several) from this exercise that does those things best. I think you'll find that none of the leads you composed will be wasted; there will be a place for the ones you don't use somewhere else in the paper. Keep them handy.

□ ■ □

Deciding on a Voice

How you begin has another subtle influence on your draft: It establishes the tone, or writing voice, you will adopt in your paper. Though you may think *writing voice* is not something you've considered much before, you probably paid a lot of attention to it when writing the essay that accompanied your college applications. Does this *sound* right? you wondered, considering whether what you wrote would impress the admissions officer. Did you sound like college material? You also know how to *change* your writing voice. For example, next time you get a speeding ticket and write home to ask for money to pay for it, notice the voice you adopt. And then, when you write your best friend a letter about the same incident, notice how your voice changes.

Of all the writing assignments you've done over the years, the research paper is probably the one in which you paid the most attention to writing voice. Research papers are supposed to sound a certain way, right? They're supposed to be peppered with words such as *myriad* and *thus* and *facilitate.* They're supposed to sound like, well, nobody you know—detached, mechanical, and ponderous.

These are understandable assumptions. So many of the sources you've read in the past weeks have sounded that way. It's also difficult to avoid sounding detached when you're writing about a topic that holds little interest for you. But the writing voice you choose for this or any other paper you write *is* a choice. Don't assume that all research papers are supposed to sound a certain way and that you must mindlessly conform to that voice.

Considering Purpose, Audience, Subject, and Who You Are

How do you choose a writing voice for a research paper? Follow the same approach you would use when writing a letter to your parents, asking for money. Consider your *purpose,* your *audience,* and your *subject.* Most importantly, though, remember that, fundamentally, your writing voice is a reflection of *who you are.* Your natural writing voice is different from mine, just as your spoken voice is. You can change your spoken voice, something you're probably pretty experienced at already. But you may need to learn to know and appreciate your written voice—the voice that sounds like you. It might even be appropriate for this paper.

I faced a difficult decision about voice in writing this text. My purpose was to instruct students in research skills as well as to motivate them to find some enthusiasm for the assignment. In order to motivate my readers, I wanted to present the research paper in a new way. That purpose would not be served, I thought, by writing in the detached, impersonal voice most people associate with textbooks (and research papers). I chose to sound like *me,* hoping that when explained in my voice, the subject would seem more accessible and my own enthusiasm for research would come through.

The Differing Voices of Research. The voice in a piece of writing often comes through in the very first line. In case you still think all research papers sound alike, listen to these first lines from student papers:

> Ernst Pawel has said that "The Metamorphosis" by Franz Kafka "transcends the standard categories of literary criticism; it is a poisoned fairy tale about the magic of hate and the power of hypocrisy . . . charting the transmogrification of a lost soul in a dead bug" (279).

—From a paper on how Kafka writes the story to deal with his own childhood demons

> If the rising waters of the Renaissance lifted the intellectual culture of Europe, it's fair to ask whether women were in the boat.

—From a paper on women artists in the Renaissance

> Even the sound of the word is vulgar.

—From a paper on ticks

> Living during a period of war was something I had never experienced until the escalation of the recent Gulf crisis.

—From a paper on Igor Stravinsky's "The Soldier's Tale"

> I have often worried in the past months if there was either something wrong with or missing from my brain.

—From a paper on dream interpretation

> No more fat jokes.

—From a paper on a daughter coming to terms with her mother's cancer

These *are* different beginnings. But notice something all these beginnings share: They are concrete. None begins with a bland, broad stroke—some sweeping generalization or obvious statement (e.g., "War is an unhappy reality in today's society" or "Richard Wright's *Native Son* is about the African-American experience in America"). Rather, each gives the reader a specific handle on the topic. In some cases, the reader is given not only a concrete point of view but, through a distinctive voice, an individual writer, as well.

The voices in the previous examples could be considered along a continuum, beginning with the more formal and moving to the much less formal, ranging from the impersonal to the personal, from a less visible writer to one who steps forward immediately. Any one of these voices might be appropriate for your paper, depending on your subject, purpose, and audience and on who you are.

Generally, as the treatment of a topic becomes more technical and its audience, more knowledgeable, the individual voice of the writer becomes less important. The writing often has less life, but then, it's not meant to entertain.

If you're writing a research paper that's intended to report the results of an experiment, you may choose a more impersonal writing voice. In such a case, it doesn't matter who you are, so don't draw attention to it. What does matter is communicating the results, simply and clearly, in a style that doesn't draw attention to itself in any way. You will find that academic writing in some disciplines is expected to assume an impersonal tone and in fact has its own language (though you may not have to strictly conform to it in this research paper).

But it's likely that at this stage, you're not writing a technical paper for an audience of experts. And though your primary purpose is not to entertain your readers, you *are* trying to make your material as interesting to others as it is to you. As suggested earlier in this book, ask your instructor if you have some latitude in choosing a voice for your paper. (See "Things to Ask Your Instructor" in the "Introduction.") If so, review the lead you tentatively chose in Exercise 4.4. Does it establish a voice that's appropriate, given your topic, purpose, and audience? Do you like the way it sounds? Should you change it? Would another lead sound better? If so, write a new lead or choose another from the several leads you wrote earlier.

Writing for Reader Interest

You've tentatively chosen a lead for your paper. You've selected it based on how well you think it frames your tentative purpose, establishes an appropriate tone or voice, and captures your readers' attention. Before you begin writing your draft, consider these other strategies for writing a lively, interesting paper that will help keep readers turning pages.

Working the Common Ground

Here's how David Quammen, a nature writer, begins an essay on the sexual strategy of Canada geese:

> Listen: *uh-whongk, uh-whongk, uh-whongk, uh-whongk,* and then you are wide awake, and you smile up at the ceiling as the calls fade off to the north and already they are gone. Silence again, 3 A.M., the hiss of March winds. A thought crosses your mind before you roll over and, contentedly, resume sleeping. The thought is: "Thank God I live here, right here exactly, in their path. Thank God for those birds." The honk of wild Canada geese passing overhead in the night is a sound to freshen the human soul. The question is why.*

If you live in Puerto Rico or anywhere beyond the late-night call of geese flying overhead, this lead paragraph may not draw you into Quammen's article on the birds' sexual habits. But for the many of us who know the muttering of geese overhead, suddenly the writer's question—why this is a sound "to freshen the human soul"—becomes our question, too. *We want to know what he knows because he starts with what we both know already:* the haunting sound of geese in flight.

David Quammen, like Richard Conniff in his article "Why God Created Flies" (see "Introduction"), understands the importance of working the common ground between his topic and his readers. In "The Miracle of Geese," Quammen begins with an experience that many of us know, and once he establishes that common ground, he takes us into the less familiar territory he encountered while researching Canada geese. And we willingly go. Quammen gives us a foothold on his topic that comes from our own experience with it.

Conniff does much the same thing throughout "Why God Created Flies." He often reminds us of familiar moments, like trying desperately to kill a fly without a fly swatter, and then helps us understand what it is about the fly that makes it so elusive, using information from his research. Conniff explains the nature of the fly's wraparound eye, its nervous system, and the speed of its wingbeats. He works the common ground with his readers even more in his explanations, using comparisons that we can relate to:

> The fly's wings beat 165 to 200 times a second. Although this isn't all that fast for an insect, it's more than double the wing beat of the speediest hummingbird and about twenty times

*David Quammen, *The Flight of the Iguana* (New York: Delacorte, 1988), 233.

faster than any repetitious movement the human nervous system can manage.*

Information about insect anatomy that alone might be dry and uninteresting is brought to life because it's brought into the world that we know through our own observations and experiences and with surprising comparisons that make sense.

As you draft your research paper, look for ways to work the common ground between your topic and your readers: What typically is their relationship to what you're writing about? What might they know about the topic but not have noticed? How does it touch their world? What would they want to know from their own experiences with your topic?

Scott, writing a paper about the town fire department that services the university, began by describing a frequent event in his dormitory: a false alarm. He then went on to explore why many alarms are not really so false after all. He hooked his readers by drawing on their common experience with his topic.

Some topics, like flies and geese and divorce and alcoholism, may have very real connections to the lives of your readers. Many people have swatted flies, heard geese overhead, or watched parents or friends destroy themselves with booze. As you revise your paper, look for opportunities to encourage readers to take a closer look at something about your topic they may have seen before.

Topics in Which Common Ground Is Hard to Find. Some topics don't yield common ground so directly. They may be outside the direct experiences of your readers. Scott was not a terrorist, needless to say, nor had he ever witnessed a terrorist attack or its aftermath; he safely assumed that his readers had never had direct experience with terrorism either. Yet it wasn't hard for Scott to imagine the common ground he could exploit in his research essay on antigovernment violence. That common ground was an event that is seared in all our memories: the Oklahoma City bombing. A description of the image of the ravaged federal building, with its many offices eerily exposed by the blast, was the opening paragraph of his essay and a symbol Scott returned to again and again.

Literary topics may also present a challenge in establishing common ground with readers, unless the author or work is familiar. But there are ways. When I was writing a paper on notions of manhood in Wallace Stegner's novels *The Big Rock Candy Mountain* and *Recapit-*

*Reprinted by permission from Conniff, Richard, *Spineless Wonders—Strange Tales of the Invertebrae World* (Henry Holt & Co., 1996).

ulation, I brought the idea of manhood home to my readers by describing my relationship with my own father and then comparing it to the relationship of two key characters in the books. Comparison to other more popular works that readers may know is often a way to establish some common ground.

Popular culture presented through the mass media—TV, radio, the Internet, magazines, newspapers—may be the common ground you can mine between your readers and a topic removed from their world. None of your readers were likely in Iraq during the Gulf War, but for a paper on whether "smart bombs" are so smart, the story of the destroyed "air raid shelter" in Baghdad, where several hundred civilians perished, will be familiar to many.

In writing your paper, imagine the ways in which your topic intersects with the life of a typical reader, and in that way, bring the information to life.

Putting People on the Page

Essayist E. B. White once advised that when you want to write about humankind, you should write about a human. The advice to look at the *small* to understand the *large* applies to most writing, not just the research paper.

Ideas come alive when we see how they operate in the world we live in. Beware, then, of long paragraphs with sentences that begin with phrases such as *in today's society,* where you wax on with generalization after generalization about your topic. Unless your ideas are anchored to specific cases, observations, experiences, statistics, and, especially, people, they will be reduced to abstractions and lose their power for your reader.

Using Case Studies. Strangely, research papers are often peopleless landscapes, which is one of the things that can make them so lifeless to read. Lisa wrote about theories of child development, citing studies and schools of thought about the topic yet never applying that information to a real child, her own daughter, two-year-old Rebecca. In his paper decrying the deforestation of the Amazon rain forest, Marty never gave his readers the chance to hear the voices of the Indians whose way of life is threatened.

Ultimately, what makes almost any topic matter to the writer or the reader is what difference it makes to people.

Candy's paper on child abuse and its effect on language development, for example, opened with the tragic story of Genie, who, for nearly thirteen years, was bound in her room by her father and beaten whenever she made a sound. When Genie was finally rescued, she could not speak at all. This sad story about a real girl makes the idea that child abuse affects how one speaks (the paper's thesis) anything but abstract. Candy gave her readers reason to care about what she learned about the problem by personalizing it.

Sometimes, the best personal experience to share is your own. Have you been touched by the topic? Kim's paper about the special problems of women alcoholics included anecdotes about several women gleaned from her reading, but the paper was most compelling when she talked about her own experiences with her mother's alcoholism.

Using Interviews. Interviews are another way to bring people to the page. In "Why God Created Flies," Richard Conniff brought in the voice of a bug expert, Vincent Dethier, who not only had interesting things to say about flies but who spoke with humor and enthusiasm. Heidi's paper on *Sesame Street* featured the voice of a school principal, a woman who echoed the point the paper made about the value of the program. Both research essays are filled not just with information about the topic but with people who are touched by it in some way.

As you write your paper, look for opportunities to bring people to the page. Hunt for case studies, anecdotes, and good quotes that will help your readers see how your topic affects how people think and live their lives.

Writing a Strong Ending

Readers remember beginnings and endings. We already explored what makes a strong beginning: It engages the reader's interest, it's more often specific than general, and it frames the purpose of the paper, defining for the reader where it is headed. A beginning for a research paper should also state or imply its thesis, or controlling idea.

We haven't said anything about endings yet, or "conclusions," as they are traditionally described. What's a strong ending? That depends. If you're writing a formal research paper (in some disciplines), the purpose of the conclusion is straightforward: It should summarize major findings. But if you're writing a less formal research essay, the nature of the conclusion is less prescribed. It could summarize major findings, but it could also suggest new directions worth exploring, highlight an especially important aspect of the topic, offer a rethinking of the thesis, or end the story of the search. The conclusion could be general, or it could be specific.

Endings to Avoid. The ending of your research paper could be a lot of things, and in a way, it's easier to say what it should *not* be:

- Avoid conclusions that simply restate what you've already said. This is the "kick the dead horse" conclusion some of us were taught to write in school on the assumption that our readers probably aren't smart enough to get our point, so we'd better repeat it. This approach annoys most readers who *are* smart enough to know the horse is dead.

- Avoid endings that begin with *in conclusion* or *thus*. Words such as these also signal to your reader what she already knows: that you're ending. Language such as this often begins a very general summary, which gets you into a conclusion such as the one mentioned above: dead.

- Avoid endings that don't feel like endings—that trail off onto other topics, are abrupt, or don't seem connected to what came before them. Prompting your readers to think is one thing; leaving them hanging is quite another.

In some ways, the conclusion of your research paper is the last stop on your journey; the reader has traveled far with you to get there. The most important quality of a good ending is that it should add something to the paper. If it doesn't, cut it and write a new one.

What can the ending add? It can add a further elaboration of your thesis that grows from the evidence you've presented, a discussion of solutions to a problem that has arisen from the information you've uncovered, or perhaps a final illustration or piece of evidence that drives home your point.

Christina Kerby's research essay on method acting (see Appendix A) explores the controversy over whether this approach was selfish, subverting the playwright's intentions about a character's identity and replacing it with the actor's focus on her own feelings and identity. Christina's ending, however, first transcends the debate by putting method acting in context: It is one of several tools an actor can use to tap her emotions for a role. But then Christina humorously raises the nagging question about selfishness once more: Can we accept that Juliet is not thinking about the fallen Romeo as she weeps by his side but about her dead cat Fluffy?

Here's Christina's ending:

> Acting is no longer about poise, voice quality, and diction. It is also about feeling the

```
part, about understanding the emotions that go
into playing the part, and about possessing the
skill necessary to bring those emotions to life
within the character. . . . Whether an actor uses
Stanislavski's method of physical actions to
unlock the door to her subconscious or whether she
attempts to stir up emotions from deep within
herself using Strasberg's method, the actor's goal
is to create a portrayal that is truthful. It is
possible to pick out a bad actor from a mile away,
one who does not understand the role because she
does not understand the emotions necessary to
create it. Or perhaps she simply lacks the means
of tapping into them.
     If genuine emotion is what the masses want,
method acting may be just what every star-struck
actress needs. Real tears? No problem. The
audience will never know that Juliet was not
lamenting the loss of her true love Romeo but
invoking the memory of her favorite cat Fluffy,
who died tragically in her arms.*
```

An ending, in many ways, can be approached similarly to a lead. You can conclude with an anecdote, a quotation, a description, a summary, or a profile. Go back to the discussion earlier in this chapter of types of leads for ideas about types of conclusions. The same basic guidelines apply.

Perhaps the strongest ending is one that somehow finds a way to circle back to the beginning. An example of this kind of conclusion—the "snake biting its tail" ending—is Conniff's last paragraph in "Why God Created Flies," where he returns to where he began his article: sitting in front of a beer, contemplating flies.

*Reprinted with permission of Christina B. Kerby.

Using Surprise

The research process—like the writing process—can be filled with discovery for the writer if he approaches the topic with curiosity and openness. When I began researching the *Lobster Almanac,* I was constantly surprised by things I didn't know: lobsters are bugs; it takes eight years for a lobster in Maine to grow to the familiar one-pound size; the largest lobster ever caught weighed about forty pounds and lived in a tank at a restaurant for a year, developing a fondness for the owner's wife. I could go on and on. And I did in the book, sharing unusual information with my readers on the assumption that if it surprised me, it would surprise them, too.

As you write your draft, reflect on the surprising things you discovered about your topic during your research and look for ways to weave that information into the rewrite. Later, after you have written your draft, share it with a reader and ask for her ideas about what is particularly interesting and should be further developed. For now, think about unusual specifics you may have left out.

However, don't include information, no matter how surprising or interesting, that doesn't serve your purpose. Christine's survey on the dreams of college freshmen had some fascinating findings, including some accounts of recurring dreams that really surprised her. She reluctantly decided not to say much about them, however, because they didn't really further the purpose of her paper, which was to discover what function dreams serve. On the other hand, Bob was surprised to find that some politically conservative politicians and judges actually supported decriminalization of marijuana. He decided to include more information about who they were and what they said in his revision, believing it would surprise his readers and strengthen his argument.

Considering Methods of Development

If you feel you have plenty of information and you're itching to get started writing the draft, maybe you should just follow your lead and see where it goes. Some of the best research papers I've read—and virtually all the research essays I've written—have grown organically from strong beginnings. When I write, I don't know what's going to happen until I see what I say.

Many people are not comfortable with such a free-fall approach to writing, including many professional writers. John McPhee is almost obsessed with the structure of his long research essays. He spends hours, positioning and repositioning index cards of information from his research on a corkboard, looking for the right way to organize his material. He does all this before he writes a word of the draft.

If you're more of a planner than a leaper, then one option is to develop an outline, a map that will guide you from the opening to the ending of your paper. (In fact, your instructor may require that an outline be submitted with the final paper.) I've always resisted outlines, largely because they seem to take the surprise out of writing the first draft for me. But many people are uncomfortable without having a sense of where they're going before they get there. And if the outline isn't rigid, it doesn't have to preclude productive surprises.

There are a variety of ways to approach an outline. It can be a short list of things you want to cover, each summarized in a few words or phrases. It can be a list of headings that neatly break down your topic. It can be a list of questions the paper will try to answer, in the order you suspect readers may ask them. Or it can be a list of topic sentences that may even begin paragraphs or sections in the draft. You decide how detailed the outline needs to be at this point.

But before you tackle an outline, it might be helpful first to decide on the basic design of the paper. The following sections review some very general methods of development that can be used alone or in combination, as they serve your purpose.

Narrative

Tell a story. It can be the story of someone who is affected by your topic, or it can be the story of what you've learned and how you learned it, a kind of narrative of thought that chronologically tells how your thinking about your topic evolved. Dan's research paper explored the connection between spouse abuse and alcoholism, beginning with the story of Louise, a woman who sought help from Dan while he was working at a counseling center. The paper continued using the narrative throughout, while Dan weaved in explanations gleaned from his research. Jessica began her research paper with a question about the meaning of a dream she had one night and then chronicled what she discovered about it and dream interpretation through her research and interviews.

Narratives often take a chronological structure (though not always), which makes them in some ways the easiest papers to organize. Can you build your paper around some story you can tell?

Problem-to-Solution

Begin by framing the problem the paper will explore, and then focus on one or more solutions that seem promising or intriguing. For example, Anne Marie's paper on acquaintance rape first set out to establish the severity of the problem and then focused on student education programs at several colleges that have met with some success in heightening student awareness of this kind of assault. Bob's paper posed the dilemma of widespread marijuana use and then developed legalization as a possible solution.

Papers that examine problems need not always provide solutions, however.

Cause-to-Effect or Effect-to-Cause

Causality, a primary interest of scientific researchers, can also be used to develop a research essay. Organize your paper around a cause of a problem, and then look at some effects, or vice versa. Consider beginning, for example, with a discussion of the dire effects of the removal of rain forests in Brazil, and then examine one cause—perhaps economic or political—that contributes to that problem. The key here is to avoid the temptation to examine *all* the causes and effects of a particular problem without ignoring its *complexity*. Things are seldom simple, and they are often interrelated. But pairing a cause and effect and building your paper around that can be illuminating. If your topic explores a problem, this may be a useful way to focus and organize your material.

Question-to-Answer

As mentioned earlier, all writing answers questions. You've been asked to identify the focusing question that your paper attempts to answer. Design the paper around that question and your exploration of the answers. Anne began her research paper with a question about the absence of preschool in her education. Is preschool necessary, she asked, and does it contribute to later academic success? The rest of Anne's paper explored the answer, culled from both research and interviews.

Carolyn's research essay on the Endangered Species Act (see Appendix B) begins by asking three questions: (1) Does the act benefit animals over humans? (2) Does it save creatures that have marginal value at the expense of the economy? and (3) Does it subvert property rights? Answering these questions becomes the organizing principle of her essay, as she examines each one in turn.

Question-to-answer is a quite natural method of development and can be combined with a narrative approach. How did you discover the answer to the question you pose? Be aware that this approach does not imply that you must find *the* answer. Again, things are rarely that simple or that neat. You may even note at the end of your paper that you asked the wrong question, a point that may be especially illuminating.

Known-to-Unknown or Unknown-to-Known

Chris investigated a mysterious murder that took place on the Isle of Shoals, a tiny cluster of islands off New Hampshire's coast, one hundred years ago. Based on documents he collected, he first examined what was known about the case, but the bulk of his paper discussed what remained a mystery.

You might also examine what is known and unknown about your topic, especially if it's the source of controversy. What do authorities on your topic seem to agree and disagree on? Where does controversy lie?

Simple-to-Complex

In a way, the simple-to-complex method of development is a variation of known-to-unknown. What is apparent about your topic? What is less obvious that reveals its complex nature? At first glance, Ken Kesey's book *One Flew Over the Cuckoo's Nest** is simply a powerful statement against institutional mistreatment of people who are mentally ill. But one student, Tim, found a more subtle, more disturbing misogynist theme. His paper first looked at the more obvious features of the novel but then focused on its less apparent antifeminist undercurrents.

General-to-Specific or Specific-to-General

Think of general-to-specific as the "funnel" approach, inverted or not. You start with a broad look at the topic and then funnel down to some more specific aspect of it, or you begin with a narrow look and end up with some broader view. Jenny's paper on how children acquire language began with a specific anecdote about Kalinda, who is asked to read in front of her class and stumbles over a word. The paper then moved to a more general discussion of the forces that shape language acquisition. From there, it got specific again, discussing particular theorists and their ideas.

Good research writing sometimes has that quality: expansion and contraction, almost like breathing, moving in and out, again and again.

Comparison-and-Contrast

Comparison-and-contrast is a strategy for organization you're no doubt familiar with. Depending on your purpose, it can work very well. For example, Nick wanted to understand what lessons the United States learned from the Vietnam War. The more recent con-

*Ken Kesey, *One Flew Over the Cuckoo's Nest* (New York: Signet, 1962).

flict with Iraq, an event that touched his life, seemed to be a useful comparison. How are the two wars different? he wondered. How are they similar? Another student, Linda, wrote a paper that compared and contrasted the creative processes involved in writing and photography. She discovered more similarities than she had expected.

Look for potentially revealing comparisons and contrasts, and organize your paper around them. You can deal with them separately—one and then the other—or you can alternate between the two, moving back and forth.

Combining Approaches

Remember that each of these methods of development can be used alone, but they will more likely be used in combination. Most papers move back and forth between the general and the specific, and many involve some kind of narrative. (After all, an anecdote or case study is a kind of story.) A paper on the destruction of the Brazilian rain forest may incorporate the cause-and-effect model mentioned above, but it may also find a place for comparison-and-contrast. For instance: What countries with rain forests have resisted economic pressures to cut trees? Would those approaches work in Brazil?

These strategies for organizing your paper are not meant to be formulas. Ignore them altogether if using them turns writing your first draft into a slow, mechanistic exercise. At best, these methods may give you a broad notion of how to organize an outline. Then you can fill in some of the detail—again, how much is up to you.

However you approach creating an outline at this stage, do so with your thesis in mind. Ask yourself, *What* do my readers need to know to understand my point, and *when* do they need to know it in my paper?

Writing with Sources

The need for *documentation*—that is, citing sources—distinguishes the research paper from most other kinds of writing. And let's face it: Worrying about sources can cramp your style. Many students have an understandable paranoia about plagiarism and tend, as mentioned earlier, to let the voices of their sources overwhelm their own. Students are also often distracted by technical details: Am I getting the right page number? Where exactly should this citation go? Do I need to cite this or not?

As you gain control of the material by choosing your own writing voice and clarifying your purpose in the paper, you should feel less constrained by the technical demands of documentation. The following suggestions may also help you weave reference sources into your own writing without the seams showing.

Blending Kinds of Writing and Sources

One of the wonderful things about the research essay is that it can draw on all four sources of information—reading, interviews, observation, and experience—as well as the four notetaking strategies discussed earlier—quotation, paraphrase, summary, and the writer's own analysis and commentary. Skillfully blended, these elements can make music.

Look at this paragraph from Heidi's paper on *Sesame Street:*

> There is more to this show than meets the eye, certainly. It is definitely more than just a crowd of furry animals all living together in the middle of New York City. Originally intended as an effort to educate poor, less privileged youth, Sesame Street is set in the very middle of an urban development on purpose (Hellman 52). As Jon Stone, one of the show's founders and co-producers sees it, the program couldn't be "just another escapist show set in a tree house or a badger den" (52). Instead, the recognizable environment gave something to the kids they could relate to. ". . . It had a lot more real quality to it than, say, Mister Rogers. . . . Kids say the reason they don't like Mister Rogers is that it's unbelievable," says Nancy Diamonti.*

The writing is lively here, not simply because the topic is interesting to those of us who know the program. Heidi has nicely blended her own commentary with summary, paraphrase, and quotation, all in a single paragraph. She has also been able to draw on multiple

*Used with permission of Heidi R. Dunham.

sources of information—an interview, some effective quotes from her reading, and her own observations of *Sesame Street.* We sense that the writer is *using* the information, not being used by it.

Handling Quotes. Avoid the temptation, as Heidi did, to load up your paragraphs with long or full quotes from your sources. I often see what I call "hanging quotes" in research papers. Embedded in a paragraph is a sentence or two within quotation marks. Though the passage is cited, there's no indication of who said it. Usually, the writer was uncertain about how to summarize or paraphrase or work *part* of the quotation into his own prose.

Use quotations selectively. And if you can, blend them into your own sentences, using a particularly striking or relevant part of the original source. For example, consider how quotes are used in this paragraph:

> Black Elk often spoke of the importance of the circle to American Indian culture. "You may have noticed that everything an Indian does is in a circle, and that is because the Power of the World always works in circles, and everything tries to be round. . . . The sky is round, and I have heard that the earth is round like a ball, and so are all the stars." He couldn't understand why white people lived in square houses. "It is a bad way to live, for there is not power in a square."

The quotes stand out, separate from the writer's own text. A better use of quotes is to work the same material smoothly into your own prose, doing something such as this:

> Black Elk believed the "Power of the World always works in circles," noting the roundness of the sun, the earth, and the stars. He couldn't understand why white people live in square houses: "It is a bad way to live, for there is not power in a square."

Occasionally, however, it may be useful to include a long quotation from one of your sources. A quotation that is longer than four lines should be *blocked,* or set off from the rest of the text by indenting it ten spaces from the left margin. Like the rest of the paper, a blocked quotation is also typed double spaced. For example:

> According to Robert Karen, shame is a particularly modern phenomenon. He notes that in medieval times, people pretty much let loose, and by our modern tastes, it was not a pretty sight:
>
> > Their emotional life appears to have been extraordinarily spontaneous and unrestrained. From Joahn Huizinga's <u>The Waning of the Middle Ages</u> we learn that the average European town dweller was wildly erratic and inconsistent, murderously violent when enraged, easily plunged into guilt, tears, and pleas for forgiveness, and bursting with psychological eccentricities. He ate with his hands out of a common bowl, blew his nose on his sleeve, defecated openly by the side of the road, made love, and mourned with great passion, and was relatively unconcerned about such notions as maladjustment or what others might think. . . . In post-medieval centuries what I've called situational shame spread rapidly. . . . (61)

Note that the quotation marks are dropped around a blocked quotation. In this case, only part of a paragraph was borrowed, but if you quote one or more full paragraphs, indent the first line of each *three* spaces in addition to the ten the block is indented from the left margin.

We'll examine *parenthetical references* more fully in the next section, but notice how the citation in the blocked quotation above is placed *outside* the final period. That's a unique exception to the usual rule that a parenthetical citation is enclosed *within* the period of the borrowed material's final sentence.

Handling Interview Material. The great quotes you glean from your interviews can be handled like quotations from texts. But there's a dimension to a quote from an interview that's lacking in a quote from a book: Namely, you participated in the quote's creation by asking a question, and in some cases, you were there to observe your subject saying it. This presents some new choices. When you're quoting an interview subject, should you enter your essay as a participant in the conversation, or should you stay out the way? That is, should you describe yourself asking the question? Should you describe the scene of the interview, your subject's manner of responding, or your immediate reaction to what she said? Or should you merely report *what* was said and *who* said it?

Christina's essay, "Crying Real Tears: The History and Psychology of Method Acting" (see Appendix A), makes good use of interviews. Notice how Christina writes about one of them in the middle of her essay:

> During a phone interview, I asked my acting teacher, Ed Claudio, who studied under Stella Adler, whether or not he agreed with the ideas behind method acting. I could almost see him wrinkle his nose at the other end of the connection. He described method acting as "self-indulgent," insisting that it encourages "island acting." Because of emotional recall, acting became a far more personal art, and the actor

began to move away from the script, often hiding the author's purpose and intentions under his own.*

Contrast Christina's handling of the Claudio interview with her treatment of material from an interview with Dave Pierini later in her essay:

Dave Pierini, a local Sacramento actor, pointed out, "You can be a good actor without using method, but you cannot be a good actor without at least understanding it." Actors are perhaps some of the greatest scholars of the human psyche because they devote their lives to the study and exploration of it. Aspiring artists are told to "get inside of the character's head." They are asked, "How would the character <u>feel</u>? How would the character <u>react</u>?"*

Do you think Christina's entry into her report of the first interview (with Ed Claudio) is intrusive? Or do you think it adds useful information or even livens it up? What circumstances might make this a good move? On the other hand, what might be some advantages of the writer staying out of the way and simply letting her subject speak, as Christina chooses to do in her treatment of the second interview (with Dave Pierini)?

Trusting Your Memory

One of the best ways to weave references seamlessly into your own writing is to avoid the compulsion to stop and study your sources as you're writing the draft. I remember that writing my research papers in college was typically done in stops and starts. I'd write a paragraph of the draft, then stop and reread a photocopy of an article, then write a few more sentences, and then stop again. Part of the

*Reprinted with permission of Christina B. Kerby.

problem was the meager notes I took as I collected information. I hadn't really taken possession of the material before I started writing the draft. But I also didn't trust that I'd remember what was important from my reading.

If, during the course of your research and writing so far, you've found a sense of purpose—for example, you're pretty sure your paper is going to argue for legalization of marijuana or analyze the symbolism on old gravestones on Cape Cod—then you've probably read purposefully, too. You *will* likely know what reference sources you need as you write the draft, without sputtering to a halt to remind yourself of what each says. Consult your notes and sources as you need them; otherwise, push them aside, and immerse yourself in your own writing.

CITING SOURCES

An Alternative to Colliding Footnotes

Like most people I knew back then, I took a typing class the summer between eighth grade and high school. Our instructional texts were long books with the bindings at the top, and we worked on standard Royal typewriters that were built like tanks. I got up to thirty words a minute, I think, which wasn't very good, but thanks to that class, I can still type without looking at the keyboard. The one thing I never learned, though, was how to turn the typewriter roller up a half space to type a footnote number that would neatly float above the line. In every term paper in high school, my footnotes collided with my sentences.

I'm certain that such technical difficulties were not the reason that most academic writers in the humanities and social sciences have largely abandoned the footnote method of citation for the parenthetical one, but I'm relieved, nonetheless. In the new system, borrowed material is parenthetically cited in the paper by indicating the author of the original work and the page it was taken from or the date it was published. These parenthetical citations are then explained more fully in the "Works Cited" page at the end of your paper where the sources themselves are listed.

By now, your instructor has probably told you what method of citing sources you should use: the Modern Language Association (MLA) style or the American Psychological Association (APA) style.

Most English classes use MLA. A complete guide to MLA conventions is provided in Appendix A and to APA, in Appendix B.

Before you begin writing your draft, go to Appendix A and read the section "Citing Sources in Your Essay." This will describe in some detail when and where you should put parenthetical references to borrowed material in the draft of your essay. Don't worry too much about the guidelines for preparing the final manuscript, including how to do the bibliography. You can deal with that next week.

I Hate These Theses to Pieces

Okay, here's a thesis:

```
I hate thesis statements.
```

And you wonder, What is this guy talking about now? What do you mean you hate thesis statements? *All* thesis statements? Why?

You'd be right to wonder for two reasons. First, my thesis statement about thesis statements isn't very good: It is too sweeping, it is overstated (*hate?*), and it deliberately withholds information. Its virtues, if any, are its shock value and the fact that it *is*—as any thesis must be—an assertion, or claim. Second, you're wondering why a teacher of writing would make such a claim in the first place. Doesn't most writing have a thesis, either stated or implied? Isn't writing that lacks a thesis unfocused, unclear? Doesn't a research paper, in particular, need a strong thesis?

Let me try again. Here's a thesis:

```
The thesis statement often discourages inquiry
instead of promoting it.
```

Hmmm . . . This is less overstated, and the claim is qualified in a reasonable way (*often discourages*). This thesis is also a bit more informative because it ever so briefly explains *why* I dislike thesis statements: *They often discourage inquiry.* But how do they do that? For one thing, when you arrive at a thesis statement prematurely, you risk turning the process of exploring your topic into a ritual hunt for examples that simply support what you already think. With this purpose in mind, you may suppress or ignore ideas or evidence that conflicts with the thesis—that threatens to disrupt the orderly march toward proving it is true.

Well, then, you infer, you're not saying you dislike *all* thesis statements, just those that people make up too soon and cling to compulsively.

Yes, I think so. I prefer what I would call the *found thesis,* the idea that you discover or the claim you come to *after* some exploration of a topic. This type of thesis often strikes me as more surprising (or less obvious) and more honest. It suddenly occurs to me, however, that I just discovered the term *found thesis* at this very moment, and I discovered it by starting with a conventional claim: *I hate thesis statements.* Doesn't that undermine my current thesis about thesis statements, that beginning with one can close off inquiry?

Well, yes, come to think of it.

What might we conclude from all of this discussion about the thesis that you can apply to the draft you're writing this week?

1. If you're already committed to a thesis, write it down. Then challenge yourself to write it again, making it somewhat more specific and informative and perhaps even more qualified.

2. At this stage, the most useful thesis may not be one that dictates the structure and arrangement of your draft but one that provides a focus for your thinking. Using the information you've collected, play out the truth of your idea or claim, but also invite questions about it—as I did—that may qualify or even overturn what you initially thought was true. In other words, use your draft to *test* the truthfulness of your thesis about your topic.

3. If you're still struggling to find a tentative thesis, use your draft to discover it. Then use your found thesis as the focus for the revision.

4. Your final draft *does* need to have a strong thesis, or controlling idea, around which the essay is built. The essay may ultimately attempt to *prove* the validity of the thesis, or the final essay may *explore* its implications.

Driving through the First Draft

You have an opening, a lot of material in your notes—much of it, written in your own words—and maybe an outline. You've considered some general methods of development, looked at ways to write with sources, and completed a quick course in how to cite them. Finish the week by writing through the first draft.

Writing the draft may be difficult. All writing, but especially research writing, is a recursive process. You may find sometimes that you must circle back to a step you took before, discovering a gap in your information, a new idea for a thesis statement, or a better lead or focus. Circling back may be frustrating at times, but it's natural and even a good sign: It means you're letting go of your preconceived ideas and allowing the discoveries you make *through writing* to change your mind.

A Draft Is Something the Wind Blows Through

Remember, too, that a *draft* is something the wind blows through. It's too early to worry about writing a research paper that's airtight, with no problems to solve. Too often, student writers think they have to write a perfect paper in the first draft. You can worry about plugging holes and tightening things up next week. For now, write a draft, and if you must, put a reminder on a piece of paper and post it on the wall next to your thesis statement. Look at this reminder every time you find yourself agonizing over the imperfections of your paper. The reminder should say, "It Doesn't Count."

Keep a few other things in mind while writing your first draft:

1. *Focus on your tentative thesis.* Use your thesis as a focus for your thinking in writing the draft. Use the information you've collected to test the validity of the claim you're making.

2. *Vary your sources.* Offer a variety of different sources as evidence to support your assertions. Beware of writing a single page that cites only one source.

3. *Remember your audience.* What do your readers want to know about your topic? What do they need to know to understand what you're trying to say?

4. *Write with your notes.* If you took thoughtful notes during the third week—carefully transforming another author's words into your own, flagging good quotes, and developing your own analysis—then you've already written at least some of your paper. You may only need to fine-tune the language in your notes and then plug them into your draft.

5. *Be open to surprises.* The act of writing is often full of surprises. In fact, it should be, since *writing* is *thinking* and the more you think about something, the more you're likely to see. You might get halfway

through your draft and discover the part of your topic that *really* fascinates you. Should that happen, you may have to change your thesis or throw away your outline. You may even have to reresearch your topic, at least somewhat. It's not necessarily too late to shift the purpose or focus of your paper (though you should consult your instructor before totally abandoning your topic at this point). Let your curiosity remain the engine that drives you forward.

5

□ ■ □

The Fifth Week

REVISING FOR PURPOSE

My high school girlfriend, Jan, was bright, warm hearted, and fun, and I wasn't at all sure I liked her much, at least at first. Though we had a lot in common—we both loved sunrise over Lake Michigan, bird watching, and Simon and Garfunkel—I found Jan a little intimidating, a little too much in a hurry to anoint us a solid "couple." But we stuck together for three years, and as time passed, I persuaded myself—despite lingering doubts—that I couldn't live without her. There was no way I was going to break my white-knuckled hold on that relationship. After all, I'd invested all that time.

As a writer, I used to have similar relationships with my drafts. I'd work on something very hard, finally finishing the draft. I'd know there were problems, but I'd developed such a tight relationship with my draft that the problems were hard to see. And even when I recognized some problems, the thought of making major changes seemed too risky. Did I dare ruin the things I loved about the draft? These decisions were even harder if the draft took a long time to write.

Revision doesn't necessarily mean you have to sever your relationship with your draft. It's probably too late to make a complete break with the draft and abandon your topic. However, revision does demand finding some way to step back from the draft and change your relationship with it, seeing it from the reader's perspective rather than just the writer's. Revision requires that you loosen your grip. And when you do, you may decide to shift your focus or rearrange the information. At the very least, you may discover gaps in information or sections of the draft that need more development. You will certainly need to prune sentences.

The place to begin is *purpose.* You should determine whether the purpose of your paper is clear and examine how well the information is organized around that purpose.

Presumably, by now you know the purpose of your essay. If you hadn't quite figured it out before you wrote last week's draft, I hope writing the draft helped you clarify your purpose. It did? Great. Then complete the following sentence. Remember that here, you're trying to focus on the *main* purpose of your draft. There are probably quite a few things that you attempt to do in it, but what is the most central purpose?

The main purpose of my essay on ____________________ is to

(use the appropriate word or words) *explain*, *argue*, *explore*,

describe ______________________________.

Here's how Christina filled in the blanks for her essay on method acting (see Appendix A):

The main purpose of my essay on *method acting* is to

explain, *argue*, *explore*, *describe* *the psychological aspects of method and its impact on American theater.*

The Thesis as a Tool for Revision

Purpose and *thesis* have a tight relationship. When I write an essay, I'm essentially in pursuit of a point, and not infrequently, it playfully eludes me. Just when I think I've figured out exactly what I'm trying to say, I have the nagging feeling that it's not quite right—it's too simplistic or obvious, it doesn't quite account for the evidence I've collected, or it just doesn't capture the spirit of the discoveries I've made. If a thesis is often a slippery fish, then having a strong sense of purpose helps me finally get a grip on it.

Purpose (and its sister *focus*) is a statement of intention—this is what I want to do in this piece of writing. It not only describes how I've limited the territory but what I plan to do when I'm there. That's why the words *explain*, *argue*, *explore*, and *describe* are so important. They pinpoint an *action* I'll take in the writing, and they'll move me toward particular assertions about what I see. One of these assertions will seem more important than any other, and that will be my thesis.

Maybe my tendency to see thesis statements as slippery is because I dislike encountering main points in essays that act like schoolyard bullies—they overcompensate for their insecurity by loudly announcing, "Hey, listen to me, bub, *I'm* the main point around here, and whaddya going to do about, huh?" Essays whose purpose is to argue something and take a broad and unqualified stand in favor or against a whole category of people/positions/theories/ideas can be the worst offenders. Things are rarely that simple, and when they are, they usually aren't very interesting to write about.

Just as often, I encounter thesis statements that act more like the kids who get singled out by the bullies for harassment. They are meek or bland assertions that would be easy to pick apart if they weren't so uninteresting. Here's one: *Nuclear bombs are so powerful, so fast, and so deadly that they have become the weapon of today.* There *are* elements of an assertion here; the writer points out that modern nuclear weapons are *fast, powerful,* and *deadly.* But this is such an obvious claim that it probably isn't even worth stating. The phrase *weapon of today* would seem more promising if it was explained a bit. What is it about nations or warfare *today* that makes such weapons so appealing? Is the apparent passion for fast, deadly, and powerful nuclear weapons today analogous to anything—maybe the passion for designer labels, fax machines, and fast food?

❑ *EXERCISE 5.1*
Dissecting the Fish

The main point in your research essay *may* be a straightforward argument—*Legalization of drugs will not, as some of its supporters claim, reduce violent crime*—or it may be an explanation or description of some aspect of your topic—*Method acting has revolutionized American theater.* But in either case, *use* the main point as a launching place for thinking about what you might do in the revision. Before you do anything else on your draft this week, consider doing the following:

- In a sentence or two, write down the thesis or controlling idea that emerged in your draft last week. It may have been stated or implied, or perhaps after writing the draft, you have a clearer idea of what you're trying to say. In any case, write down your thesis.

- Now generate a list of three or more questions that your thesis raises. These questions may directly challenge your assertion, or they may be questions—like those I raised earlier about the thesis about nuclear weapons—that help you further clarify or unpack what you're trying to say.

■ Next, rewrite your thesis statement at least three times. In each subsequent version, play with language or arrangement, add information, or get more specific about exactly what you're saying. For example:

1. *Method acting has had a major impact on American theater.*
2. *The method—which turned Stanislavski's original focus on external actions inward, toward the actor's own feelings—has generated controversy since the beginning.*
3. *An actor using the method may be crying tears, but whether they're real or not depends on whom you ask: the actor, who is thinking about her dead cat in the midst of a scene about a dying lover, or the writer, who didn't have a dead cat in mind when she wrote it.*

If this exercise works for you, several things will happen. Not only will you refine how you express your main point in the next draft, but you will also get guidance about how you might approach the revision—how you might reorganize it, what information you should add or cut, how you can further narrow your focus and even clarify your purpose. For example, the first version of the thesis on method acting provides the writer with little guidance about what information to *exclude* in the next draft. Aren't there lots of ways to show that method acting has had a major impact on American theater? The third version, on the other hand, is not only livelier and more interesting, it points the writer much more directly to what she should emphasize in the next draft: the conflict method acting creates over how theatrical roles are imagined, the license actors have with their material, and the ways that deception may be involved in a powerful performance using this technique.

What I'm suggesting here is this: Once you arrive at the controlling idea for your essay, it need not arrest your thinking about your topic, closing off any further discovery. A thesis is, in fact, a *tool* that will help you reopen the material you've gathered, rearrange it, and understand it in a fresh, new way.

Revision, as the word implies, means "re-seeing" or "reconceiving," trying to see what you failed to notice with the first look. That can be hard. Remember how stuck I was on that one picture of the lighthouse? I planted my feet in the sand, and the longer I stared through the camera lens, the harder it was to see the lighthouse from any other angle. It didn't matter that I didn't particularly like what I was seeing. I just wanted to take the picture.

You've spent more than four weeks researching your topic and the last few days composing your first draft. You may find that you've spent so much time staring through the lens—seeing your topic the way you chose to see it in your first draft—that doing a major revision is about as appealing as eating cold beets. How do you get the perspective to "re-see" the draft and rebuild it into a stronger paper?

Using a Reader

If you wanted to save a relationship, you might ask a friend to intervene. Then you'd get the benefit of a third-party opinion, a fresh view that could help you see what you may be too close to see.

A reader can do the same thing for your research paper draft. She will come to the draft without the entanglements that encumber the writer and provide a fresh pair of eyes through which you can see the work.

What You Need from a Reader

Your instructor may be that reader, or you might exchange drafts with someone else in class. You may already have someone whom you share your writing with—a roommate, a friend. Whomever you choose, try to find a reader who will respond honestly *and* make you want to write again.

What will be most helpful from a reader at this stage? Comments about your spelling and mechanics are not critical right now. You'll deal with those factors later. What the reader needs to point out is if the *purpose* of your paper is clear and if your thesis is convincing. Is it clear what your paper is about, what part of the topic you're focusing on? Does the information presented stay within that focus? Does the information clarify and support what you're trying to say? It would also be helpful for the reader to tell you what parts of the draft are interesting and what parts seem to drag.

❑ *EXERCISE 5.2*
Directing the Reader's Response

Though you could ask your reader for a completely open-ended reaction to your paper, the following questions might help her focus on providing comments that will help you tackle a revision:

1. After reading the draft, what would you say is the main question the paper is trying to answer or focus on?
2. In your own words, what is the main point?

3. What do you remember from the draft that is most convincing that the ideas in the paper are true? What is least convincing?
4. Where is the paper most interesting? Where does the paper drag?

How your reader responds to the first two questions will tell you a lot about how well you've succeeded in making the purpose and thesis of your paper clear. The answer to the third question may reveal how well you've *used* the information gleaned from research. The reader's response to the fourth question will give you a preliminary reading on how well you engaged her. Did you lose her anywhere? Is the paper interesting?

A reader responding to Jeff's paper titled "The Alcoholic Family" helped him discover some problems that are typical of first drafts. His paper was inspired by his girlfriend's struggles to deal with her alcoholic father. Jeff wondered if he could do anything to help. Jeff's reader was touched by those parts of the paper where he discussed his own observations of the troubled family's behavior; however, the reader was confused about Jeff's purpose. "Your lead seems to say that your paper is going to focus on how family members deal with an alcoholic parent," the reader wrote to Jeff, "but I thought your main idea was that people outside an alcoholic family can help but must be careful about it. I wanted to know more about how you now think you can help your girlfriend. What exactly do you need to be careful about?"

This wasn't an observation Jeff could have made, given how close he is to the topic and the draft. But armed with objective and specific information about what changes are needed, Jeff was ready to attack the draft.

Attacking the Draft

The controlling idea of your paper—that thesis you posted on an index card above your desk a week or more ago—is the heart of your paper and should, in some way, be connected to everything else in the draft.

Though a good reader can suddenly help you see things you've missed, she will likely not give much feedback on what you should do to fix these problems. Physically attacking the draft might help. If you typed your first draft, then doing this may feel sacrilegious—a little like writing in books. One of the difficulties with revision is that writers respect the typewritten page too much. When the draft is typed up, with all those words marching neatly down the page, it is hard to mess it up again. As pages emerge from the typewriter or printer, you can almost hear the sound of hardening concrete. Breaking the draft into pieces can free you to clearly see them and how they fit together.

❑ *EXERCISE 5.3* Cut-and-Paste Revision

Try this cut-and-paste revision exercise (a useful technique inspired by Peter Elbow and his book *Writing with Power**):

1. Make a photocopy of your first draft (one-sided pages only). Save the original; you may need it later.

2. Cut apart the photocopy of your research paper, paragraph by paragraph. (You may cut it into even smaller pieces later.) Once the draft has been completely disassembled, shuffle the paragraphs—get them wildly out of order so the original draft is just a memory.

3. Now go through the shuffled stack and find the *core paragraph,* the most important one in the whole paper. This is probably the paragraph that contains your thesis, or main point. This paragraph is the one that gets to the heart of what you're trying to say. Set it aside.

4. With your core paragraph directly in front of you, work your way through the remaining stack of paragraphs and make two new stacks: one of paragraphs that are relevant to your core and one of paragraphs that don't seem relevant, that don't seem to serve a clear purpose in developing your main idea. Be as tough as a drill sergeant as you scrutinize each scrap of paper. What you are trying to determine is whether each piece of information, each paragraph, is there for a reason. Ask yourself these questions as you examine each paragraph:
   - Does it *develop* my thesis or further the purpose of my paper, or does it seem an unnecessary tangent that could be part of another paper with a different focus?
   - Does it provide important *evidence* that supports my main point?
   - Does it *explain* something that's key to understanding what I'm trying to say?
   - Does it *illustrate* a key concept?
   - Does it help establish the *importance* of what I'm trying to say?
   - Does it raise (or answer) a *question* that I must explore, given what I'm trying to say?

*Peter Elbow, *Writing with Power* (New York: Oxford University Press, 1981).

You might find it helpful to write on the back of each relevant paragraph which of these purposes it serves. You may also discover that *some* of the information in a paragraph seems to serve your purpose, while the rest strikes you as unnecessary. Use your scissors to cut away the irrelevant material, pruning back the paragraph to include only what's essential.

5. You now have two stacks of paper scraps: those that seem to serve your purpose and those that don't. For now, set aside your "reject" pile. Put your core paragraph back into the "save" pile, and begin to reassemble a very rough draft, using what you've saved. Play with order. Try new leads, new ends, new middles. As you spread out the pieces of information before you, see if a new structure suddenly emerges. *But especially, look for gaps—places where you should add information.* Jot down ideas for material you might add on a piece of paper; then cut up the paper and splice (with tape) each idea in the appropriate place when you reassemble the draft in the next step. You may rediscover uses for information in your "reject" pile, as well. Mine that pile, if you need to.

6. As a structure begins to emerge, begin taping together the fragments of paper and splicing ideas for new information. Don't worry about transitions; you'll deal with those later. When you're done with the reconstruction, the draft should look drafty—something the wind can blow through—and may be totally unlike the version you started with.

□ ■ □

Examining the Wreckage

As you deal with the wreckage your scissors have wrought on your first draft, you might notice other problems with it. For example, you may discover that your draft has no real core paragraph, no part that is central to your point and purpose. Don't panic. Just make sure that you write one in the revision.

To your horror, you may find that your "reject" pile of paragraphs is bigger than your "save" pile. If that's the case, you won't have much left to work with. You may need to reresearch the topic (returning to the library this week to collect more information) or shift the focus of your paper. Perhaps both.

To your satisfaction, you may discover that your reconstructed draft looks familiar. You may have returned to the structure you started with in the first draft. If that's the case, it might mean your first draft worked pretty well; breaking it down and putting it back together simply confirmed that.

When Jeff cut up "The Alcoholic Family," he discovered immediately that his reader was right: Much of his paper did not seem clearly related to his point about the role outsiders can play in helping alcoholic families. His "reject" pile had paragraph after paragraph of information about the roles that alcoholic family members assume when there's a heavy drinker in the house. Jeff asked himself, What does that information have to do with the roles of outsiders? He considered changing his thesis, rewriting his core paragraph to say something about how each family member plays a role in dealing with the drinker. But Jeff's purpose in writing the paper was to discover what *he* could do to help.

As Jeff played with the pieces of his draft, he began to see two things. First of all, he realized that some of the ways members behave in an alcoholic family make them resistant to outside help; this insight allowed him to salvage some information from his "reject" pile by more clearly connecting the information to his main point. Second, Jeff knew he had to go back to the well: He needed to return to the library and recheck his sources to find more information on what family friends can do to help.

REVISING FOR INFORMATION

I know. You thought you were done digging. But as I said last week, research is a recursive process. (Remember, the word is *research,* or "look again.") You will often find yourself circling back to the earlier steps as you get a clearer sense of where you want to go.

As you stand back from your draft, looking again at how well your research paper accomplishes your purpose, you'll likely see holes in the information. They may seem more like craters. Jeff discovered he had to reresearch his topic, returning to the library to hunt for new sources to help him develop his point. Since he had enough time, he repeated some of the research steps from the third week, beginning with a first-level search (see "First-Level Searching" in Chapter 3). This time, though, he knew exactly what he needed to find.

You may find that you basically have what information you need but that your draft requires more development. Candy's draft on how child abuse affects language included material from some useful studies from the *Journal of Speech and Hearing Disorders,* which showed pretty conclusively that abuse cripples children's abilities to converse. At her reader's suggestion, Candy decided it was important to write more in her revision about what was learned from the studies, since they offered convincing evidence for her thesis. Though she could mine her notes for more information, Candy decided to recheck the journal indexes to look for any similar studies she may have missed. As you begin to see exactly what information you need, don't rule out another trip to the library, even this late in the game.

Finding Quick Facts

The holes of information in your research paper draft may not be large at all. What's missing may be an important but discrete fact that would really help your readers understand the point you're making. For example, in Janabeth's draft on the impact of divorce on father/daughter relationships, she realized she was missing an important fact: the number of marriages that end in divorce in the United States. This single piece of information could help establish the significance of the problem she was writing about. Janabeth could search her sources for the answer, but there's a quicker way: fact books.

Fact books are references loaded with specific, factual information that can answer most any question that is statistical in nature. Here are several key sources in which you can find fast answers to questions of fact. All are published annually, so the information is likely to be current:

PRINT SOURCES FOR QUICK FACTS AND STATISTICS

Statistical Abstract of the United States. U.S. Bureau of the Census. Washington, DC: GPO. *An astonishingly wide range of economic, social, and political statistics, culled largely from government sources. A gold mine!* (see Figure 5.1).

World Almanac and Book of Facts: 1996. Mahwah, NJ: Funk & Wagnalls. *Published since 1868, so it's useful for historical information, as well.*

Information Please Almanac, Atlas and Yearbook: 1996. 49th edition. Boston: Houghton Mifflin. *In addition to statistical information, contains material on popular culture, as well* (see Figure 5.2).

Facts on File: A Weekly World News Digest. New York: Facts on File. *Published twice monthly, a useful source of information on dates of events, names of people involved, and so on. Extremely timely information.*

ONLINE SOURCES FOR QUICK FACTS AND STATISTICS

Statistical Abstract:
http://www.census.gov/stat_abstract/
A convenient Web version; it's not quite as extensive as the bound version but features the most popular tables.

No. 133. Death Rates From Accidents and Violence: 1990 to 1992

[**Rates are per 100,000 population.** Excludes deaths of nonresidents of the United States. Beginning in 1980, deaths classified according to the ninth revision of the *International Classification of Diseases.* For earlier years, classified according to the revisions in use at the time; see text, section 2. See Appendix III]

CAUSE OF DEATH AND AGE	WHITE						BLACK					
	Male			Female			Male			Female		
	1990	1991	1992	1990	1991	1992	1990	1991	1992	1990	1991	1992
Total[1]	**81.2**	**78.7**	**76.2**	**32.1**	**31.5**	**30.4**	**142.0**	**143.9**	**134.5**	**38.6**	**38.4**	**36.7**
Motor vehicle accidents	26.1	24.4	22.4	11.4	10.8	10.2	28.1	25.6	24.0	9.4	8.7	8.8
All other accidents	23.6	23.3	23.4	12.4	12.6	12.4	32.7	34.2	30.9	13.4	13.5	12.7
Suicide	22.0	21.7	21.2	5.3	5.2	5.1	12.0	12.1	12.0	2.3	1.9	2.0
Homicide	9.0	9.3	9.1	2.8	3.0	2.8	69.2	72.0	67.5	13.5	14.2	13.1
15 to 24 years old	107.3	104.2	97.4	30.5	31.2	28.4	208.0	231.9	222.2	34.9	37.0	34.9
25 to 34 years old	97.4	94.2	90.4	26.0	24.7	23.6	218.1	213.8	193.6	48.1	47.7	46.1
35 to 44 years old	82.3	78.5	80.3	24.4	23.5	23.5	176.6	171.8	159.8	38.5	40.0	38.9
45 to 54 years old	73.5	72.9	69.8	25.3	25.2	24.2	138.5	132.4	132.9	30.7	33.1	29.5
55 to 64 years old	79.5	75.6	73.0	29.4	26.6	25.7	129.9	124.7	118.7	36.1	32.5	31.8
65 years old and over	150.7	147.4	145.4	80.1	79.6	78.6	175.5	182.2	165.7	81.6	78.6	72.6
65 to 74 years old	99.7	94.8	94.0	40.5	39.0	38.7	141.8	142.6	130.8	50.4	48.1	43.9
75 to 84 years old	195.7	190.5	186.2	89.4	87.1	83.9	206.1	213.5	201.4	95.8	89.7	89.5
85 years old and over	428.3	433.3	421.4	232.4	234.8	234.2	359.1	373.9	340.0	213.0	209.8	178.4

[1] Includes persons under 15 years old, not shown separately.

No. 134. Deaths and Death Rates From Accidents, by Type: 1970 to 1992

[See headnote, table 133 and Appendix III. See also *Historical Statistics, Colonial Times to 1970,* series B 163-165]

TYPE OF ACCIDENT	DEATHS (number)					RATE PER 100,000 POPULATION				
	1970	1980	1990	1991	1992	1970	1980	1990	1991	1992
Total	**114,638**	**105,718**	**91,983**	**89,347**	**86,777**	**56.4**	**46.7**	**37.0**	**35.4**	**34.0**
Motor vehicle accidents	54,633	53,172	46,814	43,536	40,982	26.9	23.5	18.8	17.3	16.1
Traffic	53,493	51,930	45,827	42,621	39,985	26.3	22.9	18.4	16.9	15.7
Nontraffic	1,140	1,242	987	915	997	0.6	0.5	0.4	0.4	0.4
Water-transport accidents	1,651	1,429	923	851	837	0.8	0.6	0.4	0.3	0.3
Air and space transport accidents	1,612	1,494	941	1,000	1,094	0.8	0.7	0.4	0.4	0.4
Railway accidents	852	632	663	651	642	0.4	0.3	0.3	0.3	0.3
Accidental falls	16,926	13,294	12,313	12,662	12,646	8.3	5.9	5.0	5.0	5.0
Fall from one level to another	4,798	3,743	3,194	3,291	3,091	2.4	1.7	1.3	1.3	1.2
Fall on the same level	828	415	499	474	483	0.4	0.2	0.2	0.2	0.2
Fracture, cause unspecified, and other unspecified falls	11,300	9,136	8,620	8,897	9,072	5.6	4.0	3.5	3.5	3.6
Accidental drowning	6,391	6,043	3,979	3,967	3,524	3.1	2.7	1.6	1.6	1.4
Accidents caused by—										
Fires and flames	6,718	5,822	4,175	4,120	3,958	3.3	2.6	1.7	1.6	1.6
Firearms	2,406	1,955	1,416	1,441	1,409	1.2	0.9	0.6	0.6	0.6
Electric current	1,140	1,095	670	626	525	0.6	0.5	0.3	0.2	0.2
Accidental poisoning by—										
Drugs and medicines	2,505	2,492	4,506	5,215	5,951	1.2	1.1	1.8	2.1	2.3
Other solid and liquid substances	1,174	597	549	483	498	0.6	0.3	0.2	0.2	0.2
Gases and vapors	1,620	1,242	748	736	633	0.8	0.5	0.3	0.3	0.2
Complications due to medical procedures	3,581	2,437	2,669	2,473	2,669	1.8	1.1	1.1	1.0	1.0
Inhalation and ingestion of objects	2,753	3,249	3,303	3,240	3,128	1.4	1.5	1.3	1.3	1.2

No. 135. Suicides, by Sex and Method Used: 1970 to 1992

[Excludes deaths of nonresidents of the United States. Beginning 1979, deaths classified according to the ninth revision of the *International Classification of Diseases.* For earlier years, classified according to the revision in use at the time; see text, section 2. See also *Historical Statistics, Colonial Times to 1970,* series H 979-986]

METHOD	MALE						FEMALE					
	1970	1980	1985	1990	1991	1992	1970	1980	1985	1990	1991	1992
Total	**16,629**	**20,505**	**23,145**	**24,724**	**24,769**	**24,457**	**6,851**	**6,364**	**6,308**	**6,182**	**6,041**	**6,027**
Firearms[1]	9,704	12,937	14,809	16,285	16,120	15,802	2,068	2,459	2,554	2,600	2,406	2,367
Percent of total	58	63	64	66	65	65	30	39	41	42	40	39
Poisoning[2]	3,229	2,997	3,319	3,221	3,316	3,262	3,285	2,456	2,385	2,203	2,228	2,233
Hanging and strangulation[3]	2,422	2,997	3,532	3,688	3,751	3,822	831	694	732	756	810	856
Other[4]	1,204	1,574	1,485	1,530	1,582	1,571	667	755	637	623	597	571

[1] Includes explosives in 1970. [2] Includes solids, liquids, and gases. [3] Includes suffocation. [4] Beginning 1980, includes explosives.

Source of tables 133-135: U.S. National Center for Health Statistics, *Vital Statistics of the United States,* annual; and unpublished data.

FIGURE 5.1 *Statistical Abstract of the United States* is a rich source of facts on economic, political, and social subjects.

Average Hours of Household TV Usage
(in hours and minutes per day)

	Yearly average	February	July
1985–86	7 h 10 min	7 h 48 min	6 h 37 min
1986–87	7 h 05 min	7 h 35 min	6 h 32 min
1987–88	6 h 55 min	7 h 38 min	6 h 31 min
1988–89	7 h 02 min	7 h 32 min	6 h 27 min
1989–90	6 h 55 min	7 h 16 min	6 h 24 min
1990–91	6 h 56 min	7 h 30 min	6 h 26 min
1991–92	7 h 04 min	7 h 32 min	6 h 39 min
1992–93	7 h 17 min	7 h 41 min	6 h 47 min
1993–94	7 h 21 min	7 h 51 min	6 h 53 min

Source: Nielsen Media Research, copyright 1995, Nielsen Media Research.

Weekly TV Viewing by Age
(in hours and minutes)

	Time per week	
	Nov. 1994	Nov. 1993
Women 18–24 years old	26 h 23 min	25 h 42 min
Women 25–54	30 h 55 min	30 h 35 min
Women 55 and over	44 h 11 min	44 h 11 min
Men 18–24	22 h 41 min	22 h 31 min
Men 25–54	27 h 13 min	28 h 04 min
Men 55 and over	38 h 38 min	38 h 28 min
Female Teens	20 h 20 min	20 h 50 min
Male Teens	21 h 59 min	21 h 10 min
Children 6–11	21 h 30 min	19 h 59 min
Children 2–5	24 h 42 min	24 h 32 min

Source: Nielsen Media Research, copyright 1995, Nielsen Media Research.

Television Network Addresses

American Broadcasting Companies (ABC)
77 W. 66th Street
New York, N.Y. 10023
Cable News Network (CNN)
100 International Blvd., ICNN Center
Atlanta, Ga. 30348
Canadian Broadcasting Corporation (CBC)
1500 Bronson Avenue
Ottawa, Ontario, Canada K1G 3J5
Columbia Broadcasting System (CBS)
51 W. 52nd Street
New York, N.Y. 10019
Fox Television
10201 W. Pico Blvd.
Los Angeles, Calif. 90064
National Broadcasting Company (NBC)
30 Rockefeller Plaza
New York, N.Y. 10020
Public Broadcasting Service (PBS)
1320 Braddock Place
Alexandra, Va. 22314
Turner Network Television (TNT)
100 International Blvd., ICNN Center
Atlanta, Ga. 30348
Westinghouse Broadcasting Company
888 7th Ave.
New York, N.Y. 10106

Television Set Ownership
(May 1995)

Homes with	Number	Percent
Color TV sets	94,446,000	99.0
B&W only	954,000	01.0
2 or more sets	62,964,000	66.0
One set	32,436,000	34.0
Cable	62,010,000	65.0
Total TV households	**95,400,000**	**98.0**

Source: Nielsen Media Research, copyright 1995, Nielsen Media Research.

Audience Composition by Selected Program Type[1]
(Average Minute Audience)

	General drama	Suspense and mystery drama	Situation comedy	Informational[2] 6–7 p.m.	Feature films	All regular network programs 7–11 p.m.
Women (18 and over)	7,610,000	10,220,000	8,882,000	6,910,000	8,810,000	8,270,000
Men (18 and over)	4,560,000	6,870,000	5,930,000	5,200,000	5,620,000	5,950,000
Teens (12–17)	1,010,000	760,000	1,610,000	320,000	1,220,000	1,180,000
Children (2–11)	1,180,000	990,000	2,620,000	540,000	1,820,000	1,640,000
Total persons (2+)	**14,360,000**	**18,840,000**	**18,990,000**	**12,980,000**	**17,480,000**	**17,040,000**

1. All figures are estimated for the period Nov. 1994. 2. Multiweekly viewing. *Source:* Nielsen Media Research, copyright 1995, Nielsen Media Research.

FIGURE 5.2 *Information Please Almanac* is one of several good sources for information on popular culture.

Source: From *1996 Information Please Almanac.* Reprinted with permission from Nielsen Media Research.

Census Bureau:
http://www.census.gov/main/www/subjects.html
This site is organized by subject and provides statistics on about sixty topics, everything from daycare to mining.

Almanac of American Politics:
http://politicsusa.com/PoliticsUSA/resources/almanac/
A Web version of the popular book on contemporary political figures, federal committees, and state-by-state political information.

History of Science, Technology, and Medicine:
http://www.asap.unimelb.edu.au/hstm/hstm_bio.htm
Biographical information on any important person in science, technology, or medicine; this Web site also features links to related figures, authors of each biography, and even institutions (if any) related to the work of a particular scientist.

Fact books and online sources can be valuable sources of information that will plug small holes in your draft. These references are especially useful during revision, when you often know exactly what fact you need. But even if you're not sure whether you can glean a useful statistic from one of these sources, they might be worth checking anyway. There's a good chance you'll find something useful.

REVISING FOR LANGUAGE

Most of my students have the impression that revision begins and ends with concerns about language—about *how* they said it rather than *what* they said. Revising for language is really a tertiary concern (though an important one), to be addressed after the writer has struggled with the purpose and design of a draft.

Once you're satisfied that your paper's purpose is clear, that it provides readers with the information they need to understand what you're trying to say, and that it is organized in a logical, interesting way, *then* focus your attention on the fine points of *how* it is written. Begin with voice.

Listening to the Voice

Listen to your paper by reading it aloud to yourself. You may find the experience a little unsettling. Most of us are not used to actively listening to our writing voices. But your readers will be listening.

As you read, ask yourself: Is this the voice you want readers to hear? Does it seem appropriate for this paper? Does it sound flat or wooden or ponderous in any places? Does it sound anything like you?

If revising your writing voice is necessary for any reason, begin at the beginning—the first line, the first paragraph—and rely on your ears. What sounds right?

You may discover that you begin with the right voice but lose it in places. That often happens when you move from anecdotal material to exposition, from telling a story to explaining research findings. To some extent, a shift in voice is inevitable when you move from one method of development to another, especially from personal material to factual material. But examine your word choices in those passages that seem to go flat. Do you sometimes shift to the dry language used by your sources? Can you rewrite that language in your own voice? When you do, you will find yourself cutting away unnecessary, vague, and pretentious language.

Rewriting in your own voice has another effect, too: It brings the writing to life. Readers respond to an individual writing voice. When I read David Quammen or Richard Conniff, he rises up from the page, like a hologram, and suddenly, I can see him as a distinct person. I also become interested in how each man sees the things he's writing about.

Avoid Sounding Glib

Beware, though, of a voice that calls more attention to itself than the substance of what you're saying. As you've no doubt learned from reading scholarly sources, much academic writing is voiceless, partly because what's important is not *who* the writer is but *what* he has to say.

Sometimes, in an attempt to sound natural, a writer will take on a folksy or overly colloquial voice, which is much worse than sounding lifeless. What impression does the following passage give you?

> The thing that really blew my mind was that marijuana use among college students had actually declined in the past ten years! I was psyched to learn that.

Ugh!

As you search for the right voice in doing your revision, look for a balance between flat, wooden prose, which sounds as if it were manufactured by a machine, and forced, flowery prose, which distracts the reader from what's most important: what you're trying to say.

Scrutinizing Paragraphs

How Well Do You Integrate Sources?

Last week, we looked at how to fluently blend quotation, paraphrase, and summary with your own analysis. When you achieve the right blend, the reader senses that you are in control, that the information serves your purpose. When the blend isn't quite right, you—your voice and your purpose—may disappear behind the information. As a result, your paper will seem roughly stitched together. That's the problem in the following paragraph from a research paper on the conflict with Iraq:

> George Bush and Saddam Hussein acted during the Gulf War in response to their own upbringing (Steinem 25). "Somehow, the Gulf War had turned into a conflict that both men carried out in ways that were the essence of their childhoods: one killing close up, the other from a distance; one lashing out against the world as if his life depended on it, the other striving to gain its approval with a victorious game" (Steinem 26). How we were treated as children has a great effect on how we treat the world. "Socialization determines the way this self-hatred is played out" (Steinem 26). Girls tend to respond differently than boys.

This information seems interesting enough, but it is so lifelessly woven into the paragraph that it's downright dull. The writer summarizes, then quotes, then summarizes, then quotes—without flagging who said what or even being selective about what's said. Parenthetical citations are scattered throughout, breaking the flow of the paragraph. And though the paragraph does seem to have some unity, there's no clear sense of the writer's purpose in sharing the information. The writer is "missing in action": no voice, no analysis or commentary, no sense of emphasis on what's important.

Now consider this rewrite of the same paragraph:

> One of the most interesting theories about how the Gulf War was waged has to do with the two men who waged it--George Bush and Saddam Hussein. In Revolution from Within, feminist Gloria Steinem argues that "both men carried out [the war] in ways that were the essence of their childhoods: one killing close up, the other from a distance; one lashing out against the world, . . . the other striving to gain its approval with a victorious game" (26). If Steinem is right, then the fate of the world is, and maybe always has been, determined by how well international leaders were treated by their parents.

Notice the difference? Attributing the quote—and the idea—to its source, Gloria Steinem, is one key difference in the revised paragraph. Suddenly, the voice has a name, and a well-known one at that. But the writer is more involved here, as well, from the very first sentence, noting that the theory seems interesting and concluding with a remark about why it may be important. The rewrite is also much less cluttered with parenthetical references.

Go over your draft, paragraph by paragraph, and look for ways to *use* the information from your research more smoothly. Be especially alert to "hanging quotes" that appear unattached to any source. Attribution is important. To anchor quotes and ideas to people or publications in your paper, use words such as *argues, observes, says, contends, believes,* and *offers* and phrases such as *according to.* Also look for ways to use quotes selectively, lifting key words or phrases and weaving them into your own writing. What can you add that highlights what you believe is significant about the information? How does it relate to your thesis and the purpose of your paper?

Is Each Paragraph Unified?

Each paragraph should be about one idea and organized around it. You probably know that already. But applying this notion is a particular problem in a research paper, where information abounds and paragraphs sometimes approach marathon length.

If any of your paragraphs are similar to that—that is, they seem to run nearly a page or more—look for ways to break them up into shorter paragraphs. Is more than one idea embedded in the long version? Are you explaining or examining more than one thing?

Also take a look at your shorter paragraphs. Do any present minor or tangential ideas that belong somewhere else? Are any of these ideas irrelevant? Should the paragraph be cut? The cut-and-paste exercise (5.3) done earlier this week may have helped you with this already.

Scrutinizing Sentences

Using Active Voice

Which of these two sentences seems more passive, more lifeless?

Steroids have been used by many high school athletes.

or

Many high school athletes use steroids.

The first version, written in the passive voice, is clearly the more limp of the two. It's not grammatically incorrect. In fact, you may have found texts written in the passive voice to be pervasive in the reading you've done for your research paper. Research writing is plagued by passive voice, and that's one of the reasons it can be so mind numbing to read.

Passive voice construction is simple: The subject of the sentence—the thing *doing the action*—becomes the thing *acted upon* by the verb. For instance:

Clarence *kicked the dog.*

versus

The dog was kicked by ***Clarence.***

Sometimes, the subject may be missing altogether, as in:

The study was released.

Who or *what* released it?

Active voice remedies the problem by pushing the subject up front in the sentence or adding the subject if he, she, or it is missing.

For example:

> ***High school athletes*** *use steroids.*

Knowing exactly who is using the drugs makes the sentence livelier.

Another tell-tale sign of passive voice is that it usually requires a *to be* verb: *is, was, are, were, am, be, being, been.* For example:

> *Alcoholism among women* ***has been*** *extensively studied.*

Search your draft for *be's,* and see if any sentences are written in the passive voice. (If you write on a computer, some word-processing programs will search for you.) To make a sentence active, replace the missing subject:

> ***Researchers*** *have extensively studied alcoholism among women.*

See the box on page 213 entitled "Active Verbs for Discussing Ideas," which was compiled by a colleague of mine, Cinthia Gannett. If you're desperate for an alternative to *says* or *argues,* this list offers 138 alternatives.

Using Strong Verbs

Though this may seem like nit-picking, you'd be amazed how much writing in the active voice can revitalize research writing. The use of strong verbs can have the same effect.

As you know, verbs make things happen. Some verbs can make the difference between a sentence that crackles and one that merely hums. Instead of this:

> *The league* ***gave*** *Roger Clemens, the Toronto Blue Jays pitcher, a $10,000 fine for arguing with an umpire.*

write this:

> *The league* ***slapped*** *Roger Clemens, the Toronto Blue Jays pitcher, with a $10,000 fine for arguing with an umpire.*

ACTIVE VERBS FOR DISCUSSING IDEAS

informs
reviews
argues
states
synthesizes
asserts
claims
answers
responds
critiques
explains
illuminates
determines
challenges
experiments
experiences
pleads
defends
rejects
reconsiders
verifies
announces
provides
formulates
qualifies
hints
repudiates
infers
marshalls
summarizes
disagrees
rationalizes
shifts
maintains
persists

protects
insists
handles
confuses
intimates
simplifies
narrates
outlines
allows
initiates
asserts
supports
compares
distinguishes
describes
assists
sees
persuades
lists
quotes
exposes
warns
believes
categorizes
disregards
tests
postulates
acknowledges
defies
accepts
emphasizes
confirms
praises
supplies
seeks

cautions
shares
convinces
declares
ratifies
analyzes
affirms
exaggerates
observes
substitutes
perceives
resolves
assaults
disputes
conflates
retorts
reconciles
complicates
urges
reads
parses
concludes
stresses
facilitates
contrasts
discusses
guides
proposes
points out
judges
enumerates
reveals
condemns
implies
reminds

confronts
regards
toys with
hypothesizes
suggests
contradicts
considers
highlights
disconfirms
admires
endorses
uncovers
hesitates
denies
refutes
assembles
demands
criticizes
negates
diminishes
shows
supplements
accepts
buttresses
relinquishes
treats
clarifies
grants
insinuates
identifies
explains
interprets
adds

Source: Reproduced with permission of Cinthia Gannett.

Varying Sentence Length

Some writers can sustain breathlessly long sentences, with multiple subordinate clauses, and not lose their readers. Joan Didion is one of those writers. Actually, she also knows enough not to do it too often. She carefully varies the lengths of her sentences, going from a breathless one to one that can be quickly inhaled and back again. For example, here is how her essay "Dreamers of the Golden Dream" begins. Notice the mix of sentence lengths.

> This is the story about love and death in the golden land, and begins with the country. The San Bernadino Valley lies only an hour east of Los Angeles by the San Bernadino Freeway but is in certain ways an alien place: not the coastal California of the subtropical twilights and the soft westerlies off the Pacific but a harsher California, haunted by the Mojave just beyond the mountains, devastated by the hot dry Santa Ana wind that comes down through the passes at 100 miles an hour and whines through the eucalyptus windbreaks and works on the nerves. October is a bad month for the wind, the month when breathing is difficult and the hills blaze up spontaneously. There has been no rain since April. Every voice seems a scream. It is the season of suicide and divorce and prickly dread, wherever the wind blows.*

The second sentence of Didion's lead is a whopper, but it works, especially since it's set among sentences that are more than half its length. Didion makes music here.

Examine your sentences. Are the long ones too long? You can usually tell if, when you read a sentence, there's no sense of emphasis or it seems to die out. Can you break an unnecessarily long sentence into several shorter ones? More common is a string of short, choppy sentences. For example:

> Babies are born extrasensitive to sounds. This unique sensitivity to all sounds does not last. By the end of the first year, they become deaf to speech sounds not a part of their native language.

This isn't horrible, but with some sentence combining, the passage will be more fluent:

*Joan Didion, *Slouching Toward Bethlehem* (New York: Pocket, 1968).

```
Though babies are born extrasensitive to sounds,
this unique sensitivity lasts only through the end
of the first year, when they become deaf to speech
sounds not a part of their native language.
```

Look for short sentences where you are repeating words or phrases and also for sentences that begin with pronouns. Experiment with sentence combining. The result will not only be more fluent prose but a sense of emphasis, a sense of the relationship between the information and your ideas about it.

Editing for Simplicity

Thoreau saw simplicity as a virtue, something that's obvious not only by the time he spent beside Walden Pond but also by the prose he penned while living there. Thoreau writes clearly and plainly.

Somewhere, many of us got the idea that simplicity in writing is a vice—that the long word is better than the short word, that the complex phrase is superior to the simple one. The misconception is that to write simply is to be simple minded. Research papers, especially, suffer from this mistaken notion. They are often filled with what writer William Zinsser calls *clutter.*

❑ *EXERCISE 5.4* Cutting Clutter

The following passage is an example of cluttered writing at its best (worst?). It contains phrases and words that often appear in college research papers. Read the passage once. Then take a few minutes and rewrite it, cutting as many words as you can without sacrificing the meaning. Look for ways to substitute a shorter word for a longer one and to say in fewer words what is currently said in many. Try to cut the word count by half.

```
The implementation of the revised alcohol policy
in the university community is regrettable at
the present time due to the fact that the
administration has not facilitated sufficient
student input, in spite of the fact that there
```

have been attempts by the people affected by this policy to make their objections known in many instances.
(55 words)

If you found yourself getting a little ruthless as you edited this rather dead passage, it's all right. The passage needed some machete work. A stock phrase such as *due to the fact that,* which often appears in research papers, can be resurrected quite simply by using the word *because.* A fancy word such as *implementation* can be replaced with a simple one such as *start.* There's a lot of clutter like this in the previous passage.

I hope you will also see that simplifying the prose here does not make it more simple minded but simply more clear. Cluttered writing, which is often intended to impress readers, ends up turning them off.

Of course, it's easy to be ruthless editing someone else's work. Can you be equally ruthless with your own? Take a random page of your draft research paper, and cut *at least* seven words. Look at the kinds of clutter you cut away. Do you use long words when short ones will do just as well? Do you resort to stock phrases, such as *at the present time* (*now*)? Do you signal to the reader what should be obvious with phrases such as *In conclusion* or *It should be pointed out* or *It is my opinion that?*

After you study a page of your draft and see the kinds of clutter that creeps into your writing, edit the rest. The rule is to simplify, simplify, making every word count.

□ ■ □

PREPARING THE FINAL MANUSCRIPT

I wanted to title this section "Preparing the Final Draft," but it occurred to me that *draft* doesn't suggest anything final. I always call my work a draft because until it's out of my hands, it never feels finished. You may feel that way, too. You've spent five weeks on this paper—and the last few days, disassembling it and putting it back together again. How do you know when you're finally done?

For many students, the deadline dictates that: The paper is due tomorrow. But you may find that your paper really seems to be coming together in a satisfying way. You may even like it, and you're ready to prepare the final manuscript.

Considering "Reader-Friendly" Design

Later in this section, we'll discuss the format of your final draft. Research papers in some disciplines have prescribed forms. Some papers in the social sciences, for example, require an abstract, an introduction, a discussion of method, a presentation of results, and a discussion of those results. These sections are clearly defined using subheadings, making it easier for readers to examine those parts of the paper they're most interested in. You probably discovered that in your own reading of formal research. You'll likely learn the formats research papers should conform to in various disciplines as you take upper-level courses in those fields.

While you should document this paper properly, you may have some freedom to develop a format that best serves your purpose. As you consider the format of your rewrite, keep readers in mind. How can you make your paper more readable? How can you signal your plan for developing the topic and what's important?

Some visual devices might help, including:

- Subheadings
- Graphs, illustrations, tables
- Bulleted lists (like the one you're reading now)
- Block quotes
- Underlining and paragraphing for emphasis
- White space

Long, unbroken pages of text can appear to be a gray, uninviting mass to the reader. All of the devices listed help break up the text, making it more "reader friendly." Subheadings, if not overly used, can also cue your reader to significant sections of your paper and how they relate to the whole. Long quotes, those over four lines, should be blocked, or indented ten spaces (rather than the usual five spaces customary for indenting paragraphs), separating them from the rest of the text. (See Chapter 4, "Writing with Sources," for more on blocking quotes.) Bullets—dots or asterisks preceding brief items—can be used to highlight a quick list of important information. Graphs, tables, and illustrations also break up the text, but even more important, they can help clarify and explain information. (See "Placement of Tables, Charts, and Illustrations," in Appendix A.)

The format of the book you're reading is intended, in part, to make it accessible to readers. As you revise, consider how the look of your paper can make it more inviting and easily understood.

Following MLA Conventions

I've already mentioned that formal research papers in various disciplines may have prescribed formats. If your instructor expects a certain format, he has probably detailed exactly what that format should be. But in all likelihood, your essay for this class doesn't need to follow a rigid form. It will, however, probably adhere to the basic Modern Language Association (MLA) guidelines, described in detail in Appendix A. There, you'll find methods for formatting your paper and instructions for citing sources on your "Works Cited" page. You'll also find a sample paper in MLA style by Christina Kerby, "Crying Real Tears: The History and Psychology of Method Acting." The American Psychological Association (APA) guidelines for research papers, the primary alternative to MLA, is described in Appendix B. Again, you'll also find a sample paper, this one by Carolyn Nelson and entitled "The Endagered Species Act: Protecting Both Warblers and the Trotters of the World."

Proofreading Your Paper

You've spent weeks researching, writing, and revising your paper. You want to stop now. That's understandable, no matter how much you were driven by your curiosity. Before you sign off on your research paper, placing it in someone else's hands, take the time to proofread it.

I was often so glad to be done with a piece of writing that I was careless about proofreading it. That changed about ten years ago, after I submitted a portfolio of writing to complete my master's degree. I was pretty proud of it, especially an essay about dealing with my father's alcoholism. Unfortunately, I misspelled that word—*alcoholism*—every time I used it. It was pretty humiliating.

Proofreading on a Computer

Proofreading used to involve gobbing on correction fluid to cover up mistakes and then trying to line up the paper and type in the changes. If you write on a computer, you're spared from that ordeal. The text can be easily manipulated on the screen.

Software programs can also help with the job. Most word-processing programs, for example, come with "Spell-checkers." These programs don't flag problems with sentence structure or the misuse of words, but they do catch typos and consecutive repetitions of words. A "Spell-checker" is mighty handy. Learn how to use it.

Some programs will count the number of words in your sentences, alerting you to particularly long ones, and will even point out uses of passive voice. I find some of these programs irritating because they evaluate writing ability based on factors like sentence length, which may not be a measure of the quality of your work at all. But for a basic review, these programs can be extremely useful, particularly for flagging passive construction.

A lot of writers find they need to print out their paper and proofread the "hard copy." They argue that they catch more mistakes if they proofread on paper than if they proofread onscreen. It makes sense, especially if you've been staring at the screen for days. A printed copy of your paper *looks* different, and I think you see it differently, maybe with fresher eyes and attitude. You might notice things you didn't notice before. You decide for yourself how and when to proofread.

Looking Closely

You've already edited the manuscript, pruning sentences and tightening things up. Now hunt for the little errors in grammar and mechanics that you missed. Aside from misspellings (usually typos), some pretty common mistakes appear in the papers I see. For practice, see if you can catch some of them in the following exercise, where you proofread an excerpt from a student paper.

❑ *EXERCISE 5.5*
Picking Off the Lint

I have a colleague who compares proofreading to picking the lint off an outfit, which is often your final step before heading out the door. Examine the following excerpt from a student paper. Proofread it, catching as many mechanical errors as possible. Note punctuation mistakes, agreement problems, misspellings, and anything else that seems off.

```
In an important essay, Melody Graulich notes how
"rigid dichotomizing of sex roles" in most
frontier myths have "often handicapped and
confused male as well as female writers (187),"
she wonders if a "universel mythology" (198) might
emerge that is less confining for both of them. In
```

Bruce Mason, Wallace Stegner seems to experiment with this idea; acknowledgeing the power of Bo's male fantasies and Elsa's ability to teach her son to feel. It is his strenth. On the other hand, Bruces brother chet, who dies young, lost and broken, seems doomed because he lacked sufficient measure of both the feminine and masculine. He observes that Chet had "enough of the old man to spoil him, ebnough of his mother to soften him, not enough of either to save him (Big Rock, 521)."

If you did this exercise in class, compare your proofreading of this passage with that of a partner. What did each of you find?

□ ■ □

Ten Common Mistakes

The following is a list of the ten most common errors (besides misspelled words) made in research papers that should be caught in careful proofreading. A number of these errors occurred in the previous exercise.

1. Beware of commonly confused words, such as *your* instead of *you're*. Here's a list of others:

their/there	advice/advise
know/now	lay/lie
accept/except	its/it's
all ready/already	passed/past

2. Watch for possessives. Instead of *my fathers alcoholism,* the correct style is *my father's alcoholism.* Remember that if a noun ends in *s,* still add *'s*: *Tess's laughter.* If a noun is plural, just add the apostrophe: *the scientists' studies.*

3. Avoid vague pronoun references. The excerpt in Exercise 5.5 ends with the sentence beginning with *He observes that Chet . . .* Who's *he?* The sentence should read, *Bruce observes that Chet . . .* Whenever you use the pronouns *he, she, it, they,* and *their,* make sure each clearly refers to someone or something.

4. Subjects and verbs must agree. If the subject is singular, its verb must be, too:

> *The **perils** of acid rain **are** many.*

What confuses writers sometimes is the appearance of a noun that is not really the subject near the verb. Exercise 5.5 begins, for example, with this sentence:

> *In an important essay, Melody Graulich notes how "rigid dichotomizing of sex roles" in most frontier myths **have** "often handicapped and confused male as well as female writers."*

The subject here is not *frontier myths* but *rigid dichotomizing,* a singular subject. The sentence should read:

> *In an important essay, Melody Graulich notes how "rigid dichotomizing of sex roles" in most frontier myths **has** "often handicapped and confused male as well as female writers."*

The verb *has* may sound funny, but it's correct.

5. Punctuate quotes properly. Note that commas belong inside quotation marks, not outside. Periods belong inside, too. Colons and semicolons are exceptions—they belong *outside* quotation marks. Blocked quotes don't need quotation marks at all.

6. Scrutinize use of commas. Could you substitute periods or semicolons instead? If so, you may be looking at *comma splices* or *run-on sentences.* Here's an example:

> *Since 1980, the use of marijuana by college students has steadily declined, this was something of a surprise to me and my friends.*

The portion after the comma, *this was . . . ,* is another sentence. The comma should be a period, and *this* should be capitalized.

7. Make sure each parenthetical citation *precedes* the period in the sentence you're citing but *follows* the quotation mark at the end of a sentence. In MLA style, there is no comma between the author's name and page number: *(Marks 99).*

8. Use dashes correctly. Though they can be overused, dashes are a great way to break the flow of a sentence with a related bit of information. You've probably noticed I like them. In a manuscript, type dashes as *two* hyphens (- -), not one.

9. After mentioning the full name of someone in your paper, normally use her *last name* in subsequent references. For example, this is incorrect:

> *Denise Grady argues that people are genetically predisposed to obesity.* ***Denise*** *also believes that some people are "programmed to convert calories to fat."*

Unless you know Denise or for some other reason want to conceal her last name, change the second sentence to this:

> ***Grady*** *also believes that some people are "programmed to convert calories to fat."*

One exception to this is when writing about literature. It is often appropriate to refer to characters by their first names, particularly if characters share last names (as in Exercise 5.5).

10. Scrutinize use of colons and semicolons. A colon is usually used to call attention to what follows it: a list, quotation, or appositive. A colon should follow an independent clause. For example, this won't do:

> *The most troubling things about child abuse are: the effects on self-esteem and language development.*

In this case, eliminate the colon. A semicolon is often used as if it were a colon or a comma. In most cases, a semicolon should be used as a period, separating two independent clauses. The semicolon simply implies the clauses are closely related.

Using the "Search" Function

If you're writing on a computer, use the "Search" function—a feature in most word-processing programs—to help you track down consistent problems. You simply tell the computer what word or punctuation to look for, and it will locate all occurrences in the text. For example, if you want to check for comma splices, search for commas.

The cursor will stop on every comma, and you can verify if it is correct. You can also search for pronouns to locate vague references or for words (like those listed in 1 above) you commonly misuse.

Avoiding Sexist Language

One last proofreading task is to do a *man* and *he* check. Until recently, sexism wasn't an issue in language. Use of words such as *mankind* and *chairman* was acceptable; the implication was that the terms applied to both genders. At least, that's how use of the terms was defended when challenged. Critics argued that words such as *mailman* and *businessman* reinforced ideas that only men could fill these roles. Bias in language is subtle but powerful. And it's often unintentional. To avoid sending the wrong message, it's worth making the effort to avoid sexist language.

If you need to use a word with a *man* suffix, check to see if there is an alternative. *Congressperson* sounds pretty clunky, but *representative* works fine. Instead of *mankind,* why not *humanity?* Substitute *camera operator* for *cameraman*.

Also check use of pronouns. Do you use *he* or *his* in places where you mean both genders? For example:

```
The writer who cares about his topic will bring it
to life for his readers.
```

Since a lot of writers are women, this doesn't seem right. How do you solve this problem?

1. Use *his or her, he or she,* or that mutation *s/he*. For example:

```
The writer who cares about his or her topic will
bring it to life for his or her readers.
```

This is an acceptable solution, but using *his or her* repeatedly can be awkward.

2. Change the singular subject to plural. For example:

```
Writers who care about their topics will bring
them to life for their readers.
```

This version is much better and avoids discriminatory language altogether.

3. Alternate *he* or *she*, *his* or *hers* whenever you encounter an indefinite person. If you have referred to a writer as *he* on one page, make the writer *she* on the next page. Alternate throughout.

LOOKING BACK AND MOVING ON

This book began with your writing, and it also will end with it. Before you close your research notebook on this project, open it one last time and fastwrite your response to the following questions. Keep your pen moving for seven minutes.

> How was your experience writing this research paper different from that writing others? How was it the same?

When students share their fastwrites, this comment is typical: "It was easier to sit down and write this research paper than others I've written." One student last semester added, "I think it was easier because before writing the paper, I got to research something I wanted to know about and learn the answers to questions that mattered to me." If this research project was successful, you took charge of your own learning, as this student did.

Your research paper wasn't necessarily fun. Research takes time, and writing is work. Every week, you had new problems to solve. But if the questions you asked about your topic mattered, then you undoubtedly had moments, perhaps late at night in the library, when you encountered something that suddenly cracked your topic open and let the light come pouring out. The experience can be dazzling. It's even great when it's merely interesting.

What might you take away from this research paper that will prepare you for doing the next one? At the very least, I hope you've cultivated basic research skills: how to find information efficiently, how to document, how to avoid plagiarism, and how to take notes. But hopefully, you've learned more. Perhaps you've recovered a part of you that may have been left behind when you turned eleven or twelve—the curiosity that drove you to put bugs in mayonnaise jars, read about China, disassemble a transistor radio, and wonder about Mars. Curiosity is a handy thing in college. It gets you thinking. And that's the idea.

APPENDIX A

□ ■ □

Guide to MLA Style

This section contains guidelines for preparing your essay in the format recommended by the Modern Language Association, or MLA. Part One, "Citing Sources in Your Essay," will be particularly useful as you write your draft; it provides guidance on how to parenthetically cite the sources you use in the text of your essay. Part Two, "How the Essay Should Look," will help you with formatting the manuscript after you've revised it, including guidelines for margins, tables, and pagination. Part Three, "Preparing the 'Works Cited' Page," offers detailed instructions on how to prepare your bibliography at the end of your essay; this is usually one of the last steps in preparing the final manuscript. Finally, Part Four presents a sample research essay in MLA style, which will show you how it all comes together.

PART ONE: CITING SOURCES IN YOUR ESSAY

When to Cite

Before examining the details of how to use parenthetical citations, remember when you must cite sources in your paper:

1. Whenever you quote from an original source
2. Whenever you borrow ideas from an original source, even when you express them in your own words by paraphrasing or summarizing
3. Whenever you borrow factual information from a source that is *not common knowledge*

The Common Knowledge Exception

The business about *common knowledge* causes much confusion. Just what does this term mean? Basically, *common knowledge* means facts that are widely known and about which there is no controversy.

Sometimes, it's really obvious whether something is common knowledge. The fact that the Super Bowl occurs in late January and pits the winning teams from the American and National Football Conferences is common knowledge. The fact that former president Ronald Reagan was once an actor and starred in a movie with a chimpanzee is common knowledge, too. And the fact that most Americans get most of their news from television is also common knowledge, though this information is getting close to leaving the domain of common knowledge.

But what about Christine's assertion that most dreaming occurs during rapid eye movement (REM) sleep? This is an idea about which all of her sources seem to agree. Does that make it common knowledge?

It's useful to ask next, How common to whom? Experts in the topic at hand or the rest of us? As a rule, consider the knowledge of your readers. What information will not be familiar to most of your readers or may even surprise them? Which ideas might even raise skepticism? In this case, the fact about REM sleep and dreaming goes slightly beyond the knowledge of most readers, so to be safe, it should be cited. Use common sense, but when in doubt, cite.

The MLA Author/Page System

Starting in 1984, the Modern Language Association (MLA), a body that, among other things, decides documentation conventions for papers in the humanities, switched from footnotes to the author/page parenthetical citation system. The American Psychological Association (APA), a similar body for the social sciences, promotes use of the author/date system.

You will find it fairly easy to switch from one system to the other once you've learned both. Since MLA conventions are appropriate for English classes, we will focus on the author/page system in the following sections. APA standards are explained more fully in Appendix B, which includes a sample paper.

The Basics of Using Parenthetical Citation

The MLA method of in-text citation is fairly simple: As close as possible to the borrowed material, you indicate in parentheses the original source (usually, the author's name) and the page number in the work that material came from. For example, here's how you'd cite a book or article with a single author using the author/page system:

> From the very beginning of Sesame Street in 1969, kindergarten teachers discovered that incoming students who had watched the program already knew their ABCs (Chira 13).*

The parenthetical citation here tells readers two things: (1) This information about the success of *Sesame Street* does not originate with the writer but with someone named *Chira,* and (2) readers can consult the original source for further information by looking on page 13 of Chira's book or article, which is cited fully at the back of the paper in the "Works Cited." Here is what readers would find there:

Works Cited

Chira, Susan. "Sesame Street At 20: Taking Stock." New York Times 15 Nov. 1989: 13.

Here's another example of parenthetical author/page citation from another research paper. Note the differences from the previous example:

> "One thing is clear," writes Thomas Mallon, "plagiarism didn't become a truly sore point with writers until they thought of writing as their trade. . . . Suddenly his capital and identity were at stake" (3-4).

The first thing you may have noticed is that the author's last name—Mallon—was omitted from the parenthetical citation. It didn't need to be included, since it had already been mentioned in the text. *If you mention the author's name in the text of your paper, then you only need to parenthetically cite the relevant page number(s).* This citation also tells us that the quoted passage comes from two pages rather than one.

*This and the following "Works Cited" example, along with others on pp. 231 and 247, are used with permission of Heidi R. Dunham.

Placement of Citations. Place the citation as close as you can to the borrowed material, trying to avoid breaking the flow of the sentences, if possible. To avoid confusion about what's borrowed and what's not—particularly in passages longer than a sentence—mention the name of the original author *in your paper.* Note that in the next example, the writer simply cites the source at the end of the paragraph, not naming the source in the text. Doing so makes it hard for the reader to figure out whether Blager is the source of the information in the entire paragraph or just part of it:

> Though children who have been sexually abused seem to be disadvantaged in many areas, including the inability to forge lasting relationships, low self-esteem, and crippling shame, they seem advantaged in other areas. Sexually abused children seem to be more socially mature than other children of their same age group. It's a distinctly mixed blessing (Blager 994).

In the following example, notice how the ambiguity about what's borrowed and what's not is resolved by careful placement of the author's name and parenthetical citation in the text:

> Though children who have been sexually abused seem to be disadvantaged in many areas, including the inability to forge lasting relationships, low self-esteem, and crippling shame, they seem advantaged in other areas. According to Blager, sexually abused children seem to be more socially mature than other children of their same age group (994). It's a distinctly mixed blessing.

In this latter version, it's clear that Blager is the source for one sentence in the paragraph, and the writer is responsible for the rest. Generally, use an authority's last name, rather than a formal title or first name, when mentioning her in your text. Also note that the citation is placed *inside* the period of the sentence (or last sentence) that it documents. That's almost always the case, except at the end of a blocked quotation, where the parenthetical reference is placed after the period of the last sentence. The citation can also be placed near the author's name, rather than at the end of the sentence, if it doesn't unnecessarily break the flow of the sentence. For example:

> Blager (994) observes that sexually abused children tend to be more socially mature than other children of their same age group.

How to Cite When There Is No Author. Occasionally, you may encounter a source where the author is anonymous—the article doesn't have a byline, or for some reason the author hasn't been identified. This isn't unusual with pamphlets, editorials, government documents, some newspaper articles, online sources, and short filler articles in magazines. If you can't parenthetically name the author, what do you cite?

Most often, cite the title (or an abbreviated version, if the title is long) and the page number. If you choose to abbreviate the title, begin with the word under which it is alphabetized in the "Works Cited." For example:

> Simply put, public relations is "doing good and getting credit" for it (Getting Yours 3).

Here is how the publication cited above would be listed at the back of the paper:

Works Cited

Getting Yours: A Publicity and Funding Primer for Nonprofit and Voluntary Organizations. Lincoln: Contact Center, 1991.

For clarity, it's helpful to mention the original source of the borrowed material in the text of your paper. When there is no author's name, refer to the publication (or institution) you're citing or make a more general reference to the source. For example:

> An article in Cuisine magazine argues that the best way to kill a lobster is to plunge a knife between its eyes ("How to Kill" 56).

or

According to one government report, with the current minimum size limit, most lobsters end up on dinner plates before they've had a chance to reproduce ("Size at Sexual Maturity" 3-4).

How to Cite Different Works by the Same Author. Suppose you end up using several books or articles by the same author. Obviously, a parenthetical citation that merely lists the author's name and page number won't do, since it won't be clear *which* of several works the citation refers to. In this case, include the author's name, an abbreviated title (if the original is too long), and the page number. For example:

The thing that distinguishes the amateur from the experienced writer is focus; one "rides off in all directions at once," and the other finds one meaning around which everything revolves (Murray, "Write to Learn" 92).

The "Works Cited" list would show multiple works by one author as follows:

Works Cited

Murray, Donald M. Write to Learn. 3rd ed. Fort Worth: Holt, 1990.

---. A Writer Teaches Writing: A Practical Method of Teaching Composition. Boston: Houghton, 1968.

It's obvious from the parenthetical citation which of the two Murray books is the source of the information. Note that in the parenthetical reference, no punctuation separates the title and the page number but a comma follows the author's name. If Murray had been mentioned in the text of the paper, his name could have been dropped from the citation.

How to handle the "Works Cited" page is explained more fully later in this appendix, but for now, notice that the three hyphens used in the second entry are meant to signal that the author's name in this source is the same as in the preceding entry.

How to Cite Indirect Sources. Whenever you can, cite the original source for material you use. For example, if an article on television violence quotes the author of a book and you want to use the quote, try to hunt down the book. That way, you'll be certain of the accuracy of the quote and you may find some more usable information.

Sometimes, however, finding the original source is not possible. In those cases, use the term *qtd. in* to signal that you've quoted or paraphrased a quotation from a book or article that initially appeared elsewhere. In the following example, the citation signals that Bacon's quote was culled from an article by Guibroy, not Bacon's original work:

> Francis Bacon also weighed in on the dangers of imitation, observing that "it is hardly possible at once to admire an author and to go beyond him" (qtd. in Guibroy 113).

How to Cite Personal Interviews. If you mention the name of your interview subject in your text, no parenthetical citation is necessary. On the other hand, if you don't mention the subject's name, cite it in parentheses after the quote:

> Instead, the recognizable environment gave something to kids they could relate to. "And it had a lot more real quality to it than say, Mister Rogers . . . ," says one educator. "Kids say the reason they don't like Mister Rogers is that it's unbelievable" (Diamonti).

Regardless of whether you mention your subject's name, you should include a reference to the interview in the "Works Cited." In this case, the reference would look like this:

Works Cited

Diamonti, Nancy. Personal Interview. 5 November 1990.

Sample Parenthetical References for Other Sources

MLA format is pretty simple, and we've already covered some of the basic variations. You should also know five additional variations, as follow:

1. CITING AN ENTIRE WORK

If you mention the author's name in the text, no citation is necessary. The work should, however, be listed in the "Works Cited."

> Leon Edel's Henry James is considered by many to be a model biography.

2. CITING A VOLUME OF A MULTIVOLUME WORK

If you're working with one volume of a multivolume work, it's a good idea to mention which volume in the parenthetical reference. The citation below attributes the passage to the second volume, page 3, of a work by Baym and three or more other authors. The volume number always precedes the colon, which is followed by the page number:

> By the turn of the century, three authors dominated American literature: Mark Twain, Henry James, and William Dean Howells (Baym et al. 2:3).

3. CITING SEVERAL SOURCES FOR A SINGLE PASSAGE

Occasionally, a number of sources may contribute to a single passage. List them all in one parenthetical reference, separated by semicolons:

> American soccer may never achieve the popularity it enjoys in the rest of the world, an unfortunate fact that is integrally related to the nature of the game itself (Gardner 12; "Selling Soccer" 30).*

4. CITING A LITERARY WORK

Because so many literary works, particularly classics, have been reprinted in so many editions, it's useful to give readers more infor-

*Jason Pulsifer, University of New Hampshire, 1991. Used with permission.

mation about where a passage can be found in one of these editions. List the page number and then the chapter number (and any other relevant information, such as the section or volume), separated by a semicolon. Use arabic rather than roman numerals, unless your teacher instructs you otherwise:

> Izaak Walton warns that "no direction can be given to make a man of a dull capacity able to make a Flie well" (130; ch. 5).

When citing classic poems or plays, instead of page numbers, cite line numbers and other appropriate divisions (book, section, act, scene, part, etc.). Separate the information with periods. For example, (*Othello* 3.286) indicates scene 3, line 286 of Shakespeare's work.

5. CITING AN ONLINE SOURCE

Texts on CD-ROM and online sources frequently don't have page numbers. So how can you cite them parenthetically in your essay? You have several options.

- Sometimes, the documents include paragraph numbers. In these cases, use the abbreviation *par.* or *pars.*, followed by a comma and the paragraph number or numbers you're borrowing material from. For example:

> In most psychotherapeutic approaches, the personality of the therapist can have a big impact on the outcome of the therapy ("Psychotherapy," par. 1).

- Sometimes, the material has an internal structure, such as sections, parts, chapters, or volumes. If so, use the abbreviation *sec., pt., ch.,* or *vol.* (respectively), followed by the appropriate number.

- In many cases, a parenthetical citation can be avoided entirely by simply naming the source in the text of your essay. A curious reader will then find the full citation to the article on the "Works Cited" page at the back of your paper. For example:

> According to Charles Petit, the worldwide effort to determine whether frogs are disappearing will take somewhere between three to five years.

■ Finally, if you don't want to mention the source in text, parenthetically cite the author's last name (if any) or article title:

```
The worldwide effort to determine whether frogs
are disappearing will take somewhere between three
to five years (Petit).
```

PART TWO: HOW THE ESSAY SHOULD LOOK

Printing or Typing. Type your paper on white, 8½" × 11" bond paper. Avoid the erasable variety, which smudges. If you write on a computer, make sure the printer has a fresh ribbon or sufficient ink or toner. That is especially important if you have a dot-matrix printer, which can produce barely legible pages on an old ribbon.

Margins and Spacing. The old high school trick is to have big margins. That way, you can get the length without the information. Don't try that trick with this paper. Leave one-inch margins at the top, bottom, and sides of your pages. Indent paragraphs five spaces and blocked quotes ten spaces. Double-space all of the text, including blocked quotes and "Works Cited."

Title Page. Your paper doesn't need a separate title page. Begin with the first page of text. One inch below the top of the page, type your name, your instructor's name, the course number, and the date (see following). Below that, type the title, centered on the page. Begin the text of the paper below the title.

```
Karoline Ann Fox
Professor Dethier
English 401
15 December 1991
            Metamorphosis, the Exorcist,
                    and Oedipus
     Ernst Pawel has said that Franz Kafka's "The
Metamorphosis" . . .*
```

Note that every line is double spaced. The title is not underlined (unless it includes the name of a book or some other work that should be underlined) or boldfaced.

*Reprinted with permission of Karoline A. Fox.

Pagination. Make sure that every page after the first one is numbered. That's especially important with long papers. Type your last name and the page number in the upper-righthand corner, flush with the right margin: *Ballenger 3.* Don't use the abbreviation *p.* or a hyphen between your name and the number.

Placement of Tables, Charts, and Illustrations. With MLA format, papers do not have appendixes. Tables, charts, and illustrations are placed in the body of the paper, close to the text that refers to them. Number illustrations consecutively (*Table 1* or *Figure 3*), and indicate sources below them. If you use a chart or illustration from another text, give the full citation. Place any table caption above the table, flush left. Captions for illustrations or diagrams are usually placed below them.

Handling Titles. The MLA guidelines about handling titles are, as the most recent *Handbook* observes, "strict." The general rule is that the writer should capitalize the first letters of all principal words in a title, including any that follow hyphens. The exceptions include articles (*a* and *the*), prepositions (*for, of, in, to*), coordinating conjunctions (*and, or, but, for*), and the use of *to* in infinitives. These exceptions apply *only if the words appear in the middle of a title;* capitalize them if they appear at the beginning or end.

The rules for underlining a title or putting it in quotation marks, are as follows:

1. Underline the Title if it is a book, play, pamphlet, film, magazine, TV program, CD, audiocassette, newspaper, or work of art.
2. "Put the Title in Quotes" if it is an article in a newspaper, magazine, or encyclopedia; a short story; a poem; an episode of a TV program; a song; a lecture; or a chapter or essay in a book.

Here are some examples:

The Curious Researcher (Book)

English Online: The Student's Guide to the
Internet (CD)

"Once More to the Lake" (Essay)

Historic Boise: An Introduction into the
Architecture of Boise, Idaho (Book)

"Psychotherapy" (Encyclopedia article)

Idaho Statesman (Newspaper)

"One Percent Initiative Panned" (Newspaper article)

A Word about Italic Type. If you are writing your paper on a computer or word processor, you can probably produce italic type, which is slanted to the right, *like this*. Many magazines and books—including this one—use italic type to distinguish certain words and phrases, such as titles of works that otherwise would be underlined. MLA style recommends the use of underlining, not italics. You should check with your instructor to see what style he prefers.

PART THREE: PREPARING THE "WORKS CITED" PAGE

The "Works Cited" page ends the paper. (This may also be called the "References Cited" or "Sources Cited" page, depending on the nature of your sources or the preferences of your instructor.) In the old footnote system (which, by the way, is still used in some humanities disciplines), this section used to be called "Endnotes" or "Bibliography." There are also several other lists of sources that may appear at the end of a research paper. An "Annotated List of Works Cited" not only lists the sources used in the paper but includes a brief description of each. A "Works Consulted" list includes sources that may or may not have been cited in the paper but shaped your thinking. A "Content Notes" page, keyed to superscript numbers in the text of the paper, lists short commentaries or asides that are significant but not central enough to the discussion to be included in the text of the paper.

The "Works Cited" page is the workhorse of most college papers. The other source lists are used less often. "Works Cited" is essentially an alphabetical listing of all the sources you quoted, paraphrased, or summarized in your paper. If you have used MLA format for citing sources, your paper has numerous parenthetical references to authors and page numbers. The "Works Cited" page provides complete information on each source cited in the text for the reader who wants to know. (In APA format, this page is called "References" and is only slightly different in how items are listed. See Appendix B for APA guidelines.)

If you've been careful about collecting complete bibliographic information—author, title, editor, edition, volume, place, publisher, date, page numbers—then preparing your "Works Cited" page will be easy. If you've recorded that information on notecards, all you have to do is put them in alphabetical order and then transcribe them into your paper. If you've been careless about collecting that information, you may need to take a hike back to the library.

Format

Alphabetizing the List. "Works Cited" follows the text of your paper on a separate page. After you've assembled complete information about each source you've cited, put the sources in alphabetical order by the last name of the author. If the work has multiple authors, use the last name of the first listed. If the source has no author, then alphabetize it by the first key word of the title. If you're citing more than one source by a single author, you don't need to repeat the name for each source; simply place three dashes followed by a period ("- - -.") for the author's name in subsequent listings.

Indenting and Spacing. Type the first line of each entry flush left, and indent subsequent lines of that entry (if any) five spaces. Double-space between each line and each entry. For example:

Bergquist 12

Works Cited

Bergquist, Christine. Survey. Durham, NH: University of New Hampshire.

Bierman, Dick, and Oscar Winter. "Learning during Sleep: An Indirect Test of the Erasure Theory of Dreaming." Perceptual and Motor Skills 69 (1989): 139-144.

Boxer, Sarah. "Inside Our Sleeping Minds." Modern Maturity October-November 1989: 48-54.

Brook, Stephen. The Oxford Book of Dreams. New York: Oxford UP, 1983.

Chollar, Susan. "Dreamchasers." Psychology Today April 1989: 60-61.

Foulkes, David, and Wilse B. Webb. "Sleep and Dreams." Encyclopedia Britannica. 1986 ed.

Hobson, J. Allan. The Dreaming Brain. New York: Basic Books, 1988.

Hudson, Liam. Night Life: The Interpretation of Dreams. New York: St. Martin's, 1985.

Long, Michael E. "What Is This Thing Called Sleep?" National Geographic December 1987: 787-821.*

Citing Books

You usually need three pieces of information to cite a book: the name of the author or authors, the title, and the publication information. Occasionally, other information is required. The *MLA Handbook*** lists this additional information in the order it would appear in the citation. Remember, any single entry will include a few of these things, not all of them. Use whichever are relevant to the source you're citing.

1. Name of the author
2. Title of the book (or part of it)
3. Number of edition used
5. Number of volume used
6. Name of the series
7. Where published, by whom, and the date
8. Page numbers used
9. Any annotation you'd like to add

Each piece of information in a citation is followed by a period and one space (not two).

Title. As a rule, the titles of books are underlined, with the first letters of all principle words capitalized, including those in any subtitles. Titles that are not underlined are usually those of pieces found within larger works, such as poems and short stories in anthologies. These titles are set off by quotation marks. Titles of religious works (the Bible, the Koran, etc.) are neither underlined nor enclosed within quotation marks (see the guidelines in "Handling Titles," pp. 235–236).

*Reprinted with permission of Christine M. Bergquist.

**Joseph Gibaldi, *MLA Handbook for Writers of Research Papers,* 4th ed. (New York: MLA, 1995).

Edition. If a book doesn't indicate any edition number, then it's probably a first edition, a fact you don't need to cite. Look on the title page. Signal an edition like this: *2nd ed., 3rd ed.,* and so on.

Publication Place, Publisher, and Date. Look on the title page to find out who published the book. Publishers' names are usually shortened in the "Works Cited" list: for example, *St. Martin's Press, Inc.* is shortened to *St. Martin's.*

It's sometimes confusing to know what to cite about the publication place, since several cities are often a listed on the title page. Cite the first. For books published outside the United States, add the country name along with the city to avoid confusion.

The date a book is published is usually indicated on the copyright page. If several dates or several printings by the same publisher are listed, cite the original publication date. However, if the book is a revised edition, give the date of that edition. One final variation: If you're citing a book that's a reprint of an original edition, give both dates. For example:

Stegner, Wallace. Recapitulation. 1979. Lincoln: U of Nebraska P, 1986.

This book was first published in 1979 and then republished in 1986 by the University of Nebraska Press.

Page Numbers. Normally, you don't list page numbers of a book. The parenthetical reference in your paper specifies that. But if you use only part of a book—an introduction or an essay—list the appropriate page numbers following the publication date. Use periods to set off the page numbers. If the author or editor of the entire work is also the author of the introduction or essay you're citing, list her by last name only the second time you cite her. For example:

Lee, L. L., and Merrill Lewis. Preface. Women, Women Writers, and the West. Ed. Lee and Lewis. Troy: Whitston, 1980. v-ix.

Sample Book Citations

1. A BOOK BY ONE AUTHOR

Keen, Sam. Fire in the Belly. New York: Bantam, 1991.

In-Text Citation: (Keen 101)

2. A BOOK BY TWO AUTHORS

Ballenger, Bruce, and Barry Lane. Discovering the Writer Within. Cincinnati: Writer's Digest, 1988.

In-Text Citation: (Ballenger and Lane 14)

3. A BOOK WITH MORE THAN THREE AUTHORS

If a book has more than three authors, list the first and substitute the term *et al.* for the others.

Jones, Hillary et al. The Unmasking of Adam. Highland Park: Pegasus, 1992.

In-Text Citation: (Jones et al. 21-30)

4. SEVERAL BOOKS BY THE SAME AUTHOR

Baldwin, James. Tell Me How Long the Train's Been Gone. New York: Dell-Doubleday, 1968.

---. Going to Meet the Man. New York: Dell-Doubleday, 1948.

In-Text Citation: (Baldwin, Going 34)

5. A COLLECTION OR ANTHOLOGY

Crane, R. S., ed. Critics and Criticism: Ancient and Modern. Chicago: U of Chicago P, 1952.

In-Text Citation: (Crane xx)

6. A WORK IN AN ANTHOLOGY OR COLLECTION

The title of a work that is part of a collection but was originally published as a book should be underlined. Otherwise, the title of a work in a collection should be enclosed in quotation marks.

Bahktin, Mikhail. Marxism and the Philosophy of Language. The Rhetorical Tradition. Ed.

Patricia Bizzell and Bruce Herzberg. New York: St. Martin's, 1990. 928-944.

In-Text Citation: (Bahktin 929-931)

Jones, Robert F. "Welcome to Muskie Country." <u>The Ultimate Fishing Book</u>. Ed. Lee Eisenberg and DeCourcy Taylor. Boston: Houghton, 1981. 122-134.

In-Text Citation: (Jones 131)

7. AN INTRODUCTION, PREFACE, FOREWORD, OR PROLOGUE

Scott, Jerie Cobb. Foreword. <u>Writing Groups: History, Theory, and Implications</u>. By Ann Ruggles Gere. Carbondale, IL: Southern Illinois UP, 1987. ix-xi.

In-Text Citation: (Scott x-xi)

Rich, Adrienne. Introduction. <u>On Lies, Secrets, and Silence</u>. By Rich. New York: Norton, 1979. 9-18.

In-Text Citation: (Rich 12)

8. A BOOK WITH NO AUTHOR

<u>Standard College Dictionary</u>. New York: Funk & Wagnalls, 1990.

In-Text Citation: (<u>Standard College Dictionary</u> 444)

9. AN ENCYLOPEDIA

"City of Chicago." <u>Encyclopaedia Britannica</u>. 1989 ed.

In-Text Citation: ("City of Chicago" 397)

10. A BOOK WITH AN INSTITUTIONAL AUTHOR

Hospital Corporation of America. Employee Benefits Handbook. Nashville: HCA, 1990.

In-Text Citation: (Hospital Corporation of America 5-7)

11. A BOOK WITH MULTIPLE VOLUMES

Include the number of volumes in the work between the title and publication information.

Baym, Nina et al., eds. The Norton Anthology of American Literature. 3rd ed. 2 vols. New York: Norton, 1989.

In-Text Citation: (Baym et al., vol. 2)

If you use one volume of a multivolume work, indicate which one along with the page numbers followed by the total number of volumes in the work.

Anderson, Sherwood. "Mother." The Norton Anthology of American Literature. Ed. Nina Baym et al. Vol 2. New York: Norton, 1989. 1115-1131. 2 Vols.

In-Text Citation: (Anderson 1115)

12. A BOOK THAT IS NOT A FIRST EDITION

Check the title page to determine whether the book is *not* a first edition (2nd, 3rd, 4th, etc.); if no edition number is mentioned, assume it's the first. Put the edition number right after the title.

Ballenger, Bruce. The Curious Researcher. 2nd ed. Boston: Allyn and Bacon, 1997.

In-Text Citation: (Ballenger 194)

Citing the edition is only necessary for books that are *not* first editions. This includes revised editions (*Rev. ed.*) and abridged editions (*Abr. ed.*).

13. A BOOK PUBLISHED BEFORE 1900

For a book this old, it's usually unnecessary to list the publisher.

Hitchcock, Edward. <u>Religion of Geology</u>. Glasgow, 1851.

In-Text Citation: (Hitchcock 48)

14. A TRANSLATION

Montaigne, Michel de. <u>Essays</u>. Trans. J. M. Cohen. Middlesex, England: Penguin, 1958.

In-Text Citation: (Montaigne 638)

15. GOVERNMENT DOCUMENTS

Because of the enormous variety of government documents, citing them properly can be a challenge. Since most government documents do not name authors, begin an entry for such a source with the level of government (U.S. Government, State of Illinois, etc., unless it is obvious from the title), followed by the sponsoring agency, the title of the work, and the publication information. Look on the title page to determine the publisher. If it's a federal document, then the *Government Printing Office* (abbreviated *GPO*) is usually the publisher.

United States. Bureau of the Census. <u>Statistical Abstract of the United States</u>. Washington: GPO, 1990.

In-Text Citation: (United States, Bureau of the Census 79-83)

Citing Periodicals

Periodicals—magazines, newspapers, journals, and similar publications that appear regularly—are cited similarly to books but sometimes involve different information, such as date, volume, and page numbers. The *MLA Handbook* lists the information to include in a periodical citation in the order in which it should appear:

1. Name of the author
2. Article title
3. Periodical title
4. Series number or name

5. Volume number
6. Date
7. Page numbers

Author's Name. List the author(s) as you would for a book citation.

Article Title. Unlike book titles, article titles are usually enclosed in quotation marks.

Periodical Title. Underline periodical titles, dropping introductory articles (*Aegis* not *The Aegis*). If you're citing a newspaper your readers may not be familiar with, include in the title—enclosed in brackets but not underlined—the city in which it was published. For example:

MacDonald, Mary. "Local Hiker Freezes to Death."
    Foster's Daily Democrat [Dover, NH] 28 Jan.
    1992: 1.

Volume Number. Most academic journals are numbered as volumes (or occasionally feature series numbers); the volume number should be included in the citation. Popular periodicals sometimes have volume numbers, too, but these are not included in the citations. Indicate the volume number immediately after the journal's name. Omit the tag *vol.* before the number.

There is one important variation: Though most journals number their pages continuously, from the first issue every year to the last, a few don't. These journals feature an issue number as well as a volume number. In that case, cite both by listing the volume number, a period, and then the issue number: for example *12.4,* or volume number *12* and issue *4*.

Date. When citing popular periodicals, include the day, month, and year of the issue you're citing—in that order—following the periodical name. Academic journals are a little different. Since the volume number indicates when the journal was published within a given year, just indicate that year. Put it in parentheses following the volume number and before the page numbers (see examples following).

Page Numbers. Include the page numbers of the article at the end of the citation, followed by a period. Just list the pages of the entire article, omitting abbreviations such as *p.* or *pp.* It's common for articles in newspapers and popular magazines *not* to run on consecutive pages. In that case, indicate the page on which the article begins, followed by a "+" (*12+*).

Newspaper pagination can be peculiar. Some papers wed the section (usually a letter) with the page number (*A4*); other papers simply begin numbering anew in each section. Most, however, paginate continuously. See the following sample citations for newspapers for how to deal with these peculiarities.

Online sources, which often have no pagination at all, present special problems. For guidance on how to handle them, see the section "Citing Online Sources" later in this part of the Appendix.

Sample Periodical Citations

1. A MAGAZINE ARTICLE

Probasco, Steve. "Nymphing Tactics for Winter Steelhead." Fly Fisherman. February 1992: 36-39.

In-Text Citation: (Probasco 36)

Jones, Thom. "The Pugilist at Rest." New Yorker 12 Dec. 1991: 38-47.

In-Text Citation: (Jones 40)

2. A JOURNAL ARTICLE

For an article that is paginated continuously, from the first issue every year to the last, cite as follows:

Allen, Rebecca E., and J. M. Oliver. "The Effects of Child Maltreatment on Language Development." Child Abuse and Neglect 6 (1982): 299-305.

In-Text Citation: (Allen and Oliver 299-300)

For an article in a journal that begins pagination with each issue, include the issue number along with the volume number.

Goody, Michelle M., and Andrew S. Levine. "Health-Care Workers and Occupational Exposure to AIDS." Nursing Management 23.1 (1992): 59-60.

In-Text Citation: (Goody and Levine 59)

3. A NEWSPAPER ARTICLE

Some newspapers have several editions (morning edition, late edition, national edition), and each may feature different articles. If an edition is listed on the masthead, include it in the citation.

Gelbspan, Ross. "NRC Staff Lied about Seabrook, Study Says." Boston Globe 29 Jan. 1992, morning ed.: 1+.

In-Text Citation: (Gelbspan 1)

Some papers begin numbering pages anew in each section. In that case, include the section number if it's not part of pagination.

Brooks, James. "Lobsters on the Brink." Portland Press 29 Nov. 1988, sec. 2: 4.

In-Text Citation: (Brooks 4)

4. AN ARTICLE WITH NO AUTHOR

"The Understanding." New Yorker 2 Dec. 1991: 34-35.

In-Text Citation: ("Understanding" 35)

5. AN EDITORIAL

"The Star Search." Editorial. Boston Globe 29 Jan. 1992: 10.

In-Text Citation: ("Star Search" 10)

6. A LETTER TO THE EDITOR

Levinson, Evan B. "Paying Out of Pocket for Student Supplies." Letter. Boston Globe 29 Jan. 1992: 10.

In-Text Citation: (Levinson 10)

7. A REVIEW

Page, Barbara. Rev. of Allegories of Cinema: American Film in the Sixties by David E. James. College English 54 (1992): 945-954.

In-Text Citation: (Page 945-946)

8. AN ABSTRACT FROM *DISSERTATION ABSTRACTS INTERNATIONAL*

McDonald, James C. "Imitation of Models in the History of Rhetoric: Classical, Belletristic, and Current-Traditional." DAI 48 (1988): 2613A. U of Texas, Austin.

In-Text Citation: (McDonald 2613A)

Citing Nonprint and Other Sources

1. AN INTERVIEW

If you conducted the interview yourself, list your subject's name first, indicate what kind of interview it was (telephone interview, e-mail interview, or personal interview), and provide the date.

Diamonti, Nancy. Personal interview. 5 Nov. 1990.

In-Text Citations: (Diamonti)

Or avoid parenthethical reference altogether by mentioning the subject's name in the text: According to Nancy Diamonti, . . .

If you're citing an interview done by someone else (perhaps from a book or article) and the title does not indicate that it was an interview, you should, after the subject's name. Always begin the citation with the subject's name.

Stegner, Wallace. Interview. Conversations with Wallace Stegner. By Richard Eutlain and Wallace Stegner. Salt Lake: U of Utah P, 1990.

In-Text Citations: (Stegner 22)

Or if there are other works by Stegner on the "Works Cited" page: (Stegner, Conversations 22)

2. SURVEYS, QUESTIONNAIRES, AND CASE STUDIES

If you conducted the survey or case study, list it under your name and give it an appropriate title.

Bergquist, Christine. "Dream Questionnaire."
Durham: U of New Hampshire, 1990.

In-Text Citation: (Bergquist)

3. RECORDINGS

Generally, list a recording by the name of the performer and underline the title. Also include the recording company, catalog number, and year. (If you don't know the year, use the abbreviation *n.d.*)

Orff, Carl. Carmina Burana. Cond. Seiji Ozawa.
Boston Symphony. RCA, 6533-2-RG, n.d.

In-Text Citation: (Orff)

4. TELEVISION AND RADIO PROGRAMS

List the title of the program (underlined), the station, and the date. If the episode has a title, list that first in quotation marks. You may also wish to include the name of the narrator or producer after the title.

All Things Considered. Interview with Andre Dubus.
NPR. WBUR, Boston. 12 Dec. 1990.

In-Text Citation: (All Things Considered)

5. FILMS AND VIDEOTAPES

Begin with the title (underlined), followed by the director, the distributor, and the year. You may also include names of writers, performers, or producers. End with the date and any other specifics about the characteristics of the film or videotape that may be relevant (length and size).

Touchstone. Videocassette. Dir. Jim Stratton.
Alaska Conservation Foundation, 1990. 25 min.

In-Text Citation: (Touchstone)

You can also list a video or film by the name of a contributor you'd like to emphasize.

Capra, Frank, dir. It's a Wonderful Life. With Jimmy Stewart and Donna Reed.

In-Text Citation: (Capra)

6. ARTWORKS

List each work by artist. Then cite the title of the work (underlined) and where it's located (institution and city). If you've reproduced the work from a published source, include that information, as well.

Homer, Winslow. Casting for a Rise. Hirschl and Adler Galleries, New York. Illus. in Ultimate Fishing Book. Ed. Lee Eisenberg and DeCourcy Taylor. Boston: Houghton, 1981.

In-Text Citation: (Homer 113)

7. LECTURES AND SPEECHES

List each by the name of the speaker, followed by the title of the address (if any) in quotation marks, the name of the sponsoring organization, the location, and the date. Also indicate what kind of address it was (lecture, speech, etc.).

Tsongas, Paul. Speech. Tsongas for President Campaign. Durham, 28 Jan. 1992.

Avoid the need for parenthetical citation by mentioning the speaker's name in your text.

8. A PAMPHLET

Cite a pamphlet as you would a book.

New Challenges for Wilderness Conservationists. Washington, DC: Wilderness Society, 1973.

In-Text Citation: (New Challenges)

Citing CD-ROMs, Diskettes, and Magnetic Tapes

Nearly every new computer these days is sold with an encyclopedia on CD-ROM. If you're doing research, I don't think they hold a candle to the more extensive bound versions. Still, a CD-ROM encyclopedia is easy to use and, for quickly checking facts, can be quite helpful. While the encyclopedia is the most familiar *portable* database on CD-ROM, there are many others, including full-text versions of literary classics, journal article abstracts, indexes, and periodicals. The number of such portable databases on CD will continue to multiply along with databases on other media, like diskettes and tapes. Citation of these materials requires much of the usual information and in the usual order. But it will also include these three things: the *publication medium* (for example, CD-ROM, diskette, or tape), the *vendor* or company that distributed it (for example, SilverPlatter or UMI-Proquest), and the *date of electronic publication* (or the release date of the disk or tape).

There are two categories of portable databases: (1) those that are issued periodically, like magazines and journals, and (2) those that are not routinely updated, like books. Citing a source in each category requires some slightly different information.

1. A NONPERIODICAL DATABASE

This is cited much like a book.

- List the author. If no author is given, list the editor or translator, followed by the appropriate abbreviation (*ed.*, *trans.*)
- Publication title (underlined) or title of the portion of the work you're using (if relevant)
- Name of editor, compiler, or translator (if relevant)
- Publication medium (for example, CD-ROM, diskette, magnetic tape)
- Edition or release or version
- City of publication
- Publisher and year of publication

For example:

```
Shakespeare, William. Romeo and Juliet. Diskette.
     Vers. 1.5. New York: CMI, 1995.
```

In-Text Citation: (Shakespeare)

"Psychotherapy." Microsoft Encarta. 1994 ed.
CD-ROM. Everett, WA: Microsoft, 1993.

In-Text Citation: ("Psychotherapy")

2. A PERIODICAL DATABASE

Frequently a periodical database is a computer version—or an analogue—of a printed publication. For example, the *New York Times* has a disk version, as does *Dissertation Abstracts*. Both databases refer to articles also published in print, therefore, the citation often includes two dates: the original publication date and the electronic publication date. Note the location of each in the citations below.

Haden, Catherine Ann. "Talking About the Past with Preschool Siblings." DAI 56 (1996). Emory U, 1995. Dissertation Abstracts Ondisc. CD-ROM. UMI-Proquest. March 1996.

Kolata, Gina. "Research Links Writing Style to the Risk of Alzheimer's." New York Times 21 Feb. 1996: 1A. Newspaper Abstracts. CD-ROM. UMI-Proquest. 1996.*

In-Text Citation: (Kolata 1A)

Frequently, a periodically issued electronic source doesn't have a printed analogue. In that case, obviously, you can't include publication information about the printed version.

Citing Online Databases

In the first edition of *The Curious Researcher,* I barely mentioned online—or *nonportable*—databases. Since then, the amount of research information available from the Internet and commercial online services has exploded. Unfortunatley, so has confusion about how to cite such material properly. Thankfully, the Modern Language Association (MLA) published the second edition of its *MLA Style Manual*

*Sometimes information about an electronic source is unavailable. In that case, include what information you have. For example, in this example, I was unable to find the month of publication for the *Newspaper Abstracts,* and had to omit that piece of information from the citation.

in the spring of 1998. This manual clarifies and simplifies the conventions for citing online sources. The folks at MLA admit this information won't be the last word on citing online sources because electronic sources keep evolving. But the citation information in the latest *MLA Style Manual* (a book intended for graduate students and scholars) is an improvement over that in the *MLA Handbook,* fourth edition (the manual most undergraduates use to reference citing sources). You can keep up-to-date on MLA guidelines for citing online material by visiting their web site:

http://www.mla.org/main_stl.htm

Some gaps and ambiguities in the MLA treatment of online sources remain. While some competing online citation proposals from groups like the Alliance for Computers and Writing exist, I decided to cover the MLA conventions for citing online sources in *The Curious Researcher.*

Citing most online sources is much like citing any other sources, with two crucial exceptions:

1. Electronic-source citations usually include at least two dates: the *date of electronic publication* and the *date of access* (when you visited the site and retrieved the document). There is a good reason for listing both dates: online documents are changed and updated frequently—when you retrieve the material matters. If the online document you are using originally appeared in print, it might be necessary to include three dates: the print publication date, the online publication date, and your access date (see the McGrory citation in the section, "Is It in Also Print?").

2. The MLA now requires that you include the Internet address of the document in angle brackets at the end of your citation (for example, <http:www.cc.emory.edu/citation.formats.html>). The reason is obvious: the Internet address tells your readers where they can find the document.

Other Recent Changes by the MLA. The MLA no longer requires inclusion of a number of items in a citation. For example, including the word *Online* in your citations to indicate the publication medium and mentioning the name of the network or service you used to retrieve the document (for example, *Internet, America Online*) are no longer necessary. Both are great improvements, I think. Another quirky thing about citing online sources is dealing with page numbers, paragraph numbers, or numbered sections. Many Internet documents simply don't have them. The MLA no longer requires inclusion of the term *n. pag.* when a document lacks pagination.

Is It Also in Print? Databases from computer services or networks feature information available in printed form (like a newspaper or magazine) and online, or information available exclusively online. This distinction is important. If the online source has a printed version, include information about it in the citation. For example:

McGrory, Brian. "Hillary Clinton's Profile Boosted." Boston Globe Online 27 June 1996. Boston Globe 26 June 1996: 1. 8 July 1998 <http://www.boston.com/80/globe/nat/cgi-bin>.

In-Text Citation: (McGrory 1)

Note that the first date lists when the print version appeared, the second date when the article was published online, and the third when the researcher accessed the document.

Material that only appeared online is somewhat simpler to cite since you'll only need to include information about the electronic version.

Adler, Jonathan. "Save Endangered Species, Not the Endangered Species Act." The Heartland Institute: Intellectual Ammunition. Jan.-Feb. 1996. 4 Oct. 1996 <http://www.heartland.org/05jnfb96.htm>.

In-Text Citation: No page or paragraph numbers were used in this document, so simply list the author's last name: (Adler). Or avoid parenthetical citation altogether by mentioning the name of the source in your essay (for example: "According to Jonathan Adler, the ESA is . . .").

You may be missing citation information on some Internet material—like page numbers and publication dates—that are easy to find in printed texts. Use the information that you have. Keep in mind that the relevant information for a citation varies with the type of electronic source (see citation examples in the section titled "Sample Online Citations"). To summarize, the basic format for an online citation includes the following information:

1. Author's name (if given). If there is an editor, translator, or compiler included, list that name followed by the appropriate abbreviation (*ed., trans., comp.*).

2. Publication information:
   - title of the document, database, or web site
   - title of the larger work, database, or web site (if any) of which it is a part
   - name of editor (if any) of the project, database, or web site (usually different from author)
   - volume, issue, or version number (if any)
   - date of electronic publication or latest update
   - page or paragraph numbers (if any)
   - publication information about print version (if any)
   - date of access and electronic address

Address Mistakes Are Fatal. When you include Internet addresses in your citations, it is crucial that you take great care in accurately recording them. Make sure you get all your slash marks going in the right direction and the right characters in the right places; also pay attention to whether the characters are upper- or lowercase. These addresses are *case sensitive*, unlike, say, the file names used to retrieve WordPerfect documents. The cut-and-paste function in your word processor is an invaluable tool in accurately transferring Internet addresses into your own documents. One last thing: if an Internet address in your citation must go beyond one line, make sure the break occurs after a slash, not in the middle of a file name, and don't include an end-of-line hyphen to mark the break.

Sample Online Citations

1. AN ARTICLE IN AN ONLINE JOURNAL, MAGAZINE, OR NEWSPAPER

Notice the inclusion of the document length after the publication date in these examples. Sometimes Internet documents number paragraphs instead of page numbers. Include that information, if available, using the abbreviation *par.* or *pars.* (e.g., "53 pars."). More often, an Internet article has no page or paragraph numbers. Put the title of the article in quotation marks and underline the title of the journal, newsletter, or electronic conference.

Haynes, Cynthia and Jan R. Holmevik. "Enhancing Pedagogical Reality with MOOs." Kairos: A Journal for Teachers of Writing in a Webbed Environment 1.2 (1996): 1 pg. 28 June 1996 <http://english/ttu.edu/kairos/1.2/index.html>.

In-Text Citation: (Haynes and Holmevik 1)

"Freeman Trial Delayed over Illness."
USA Today. 26 May 1998. 26 May 1998
<http://www.usatoday.com/news/nds2.htm>.

In-Text Citation: ("Freeman")

Dvorak, John C. "Worst Case Scenarios." PC
Magazine Online. 26 May 1998: 3 pgs. 1 June
1998 <http://www.zdnet.com/pcmag/insites/
dvorak/jd.htm>.

In-Text Citation: (Dvorak 2)

2. AN ONLINE BOOK

I can't imagine why anyone would read *The Adventures of Huckleberry Finn* online, but it's available, along with thousands of other books and historical documents in electronic form. If you use an online book, remember to include publication information (if available) about the original printed version in the citation.

Twain, Mark. The Adventures of Huckleberry Finn.
New York: Harper, 1912. 22 July 1996
<gopher://wiretap.spies.com/00/Library/
Classic/huckfinn.html>.

In-Text Citation: (Twain) Or better yet, since there are no page numbers, mention the author in the text rather than citing him parenthetically: In The Adventures of Huckleberry Finn, Twain re-creates southern dialect . . .

When citing part of a larger work, include the title of that smaller part in quotation marks before the title of the work. Also notice that the text cited below is part of an online scholarly project. Include the name of the project, the editor and compiler of the work if listed, and its location.

Service, Robert. "The Mourners." Rhymes of a
Red Cross Man. 1916. Project Gutenberg.
Ed. A. Light. Aug. 1995. Illinois

Benedictine College. 1 July 1998 <ftp://uiarchive.cso.uiuc.edu/pub/etext/gutenberg/etext95/redcr10.txt>.

In-Text Citation: (Service)

3. A PERSONAL OR PROFESSIONAL WEB SITE

Begin with the name of the editor or creator of the site, if listed. Include the title of the site, or, if no title is given, use a descriptor such as the term *Home page*. Also include the sponsoring organization, if any, the date of access, and the electronic address.

Sharev, Alexi. Population Ecology. Virginia Tech U. 7 Aug. 1998 <http://www.gypsymoth.ento.vt.edu/~sharov/popechome/welcome.html>.

In-Text Citation: (Sharev)

Battalio, John. Home page. 26 May 1998 <http://www.idbsu.edu/english/jbattali>.

In-Text Citation: (Battalio)

You may cite a document that is part of a web site. For example:

Cohn, Priscilla. "Wildlife Contraception: An Introduction." Animal Rights Law Center Web Site. 1998. Rutgers U. 27 May 1998 <http://www.animal-law.org/hunting/contintro.htm>.

In-Text Citation: (Cohn)

4. AN ONLINE POSTING

An online post can be a contribution to an e-mail discussion group like a listserv, a post to a bulletin board or Usenet group, or a WWW forum. The description *Online posting* is included after the title of the message (usually drawn from the subject line). List the date the material was posted, the access date, and the online address as you would for any other online citation.

Alvoeiro, Jorge. "Neurological Effects of Music." Online posting. 20 June 1996. 10 Aug. 1996 <news:sci.psychology.misc>.

In-Text Citation: (Alvoerio)

The following example is from an e-mail discussion group. The address at the end of the citation is from the group's archives, available on the Web. If you don't have an Internet address for the post you want to cite, include the e-mail address of the group's moderator or supervisor.

Ledgerberg, Joshua. "Re: You Shall Know Them." Online posting. 2 May 1997. Darwin Discussion Group. 27 May 1998 <http://rjohara.uncg.edu>.

In-Text Citation: (Ledgerberg)

5. AN E-MAIL MESSAGE

Tobin, Lad. "Teaching the TA Seminar." E-mail to the author. 8 July 1996.

In-Text Citation: (Tobin)

6. A SOUND CLIP

Gonzales, Richard. "Asian American Political Strength." Natl. Public Radio. 27 May 1998. 12 July 1998 <http://www.npr.org/ramfiles/980527.me.12.ram>.

In-Text Citation: (Gonzales)

7. AN INTERVIEW

Boukreev, Anatoli. Interview. Outside Online 14 Nov. 1997. 27 May 1998 <http://outside.starwave.com/news/123097/anatolitrans.html>.

In-Text Citation: (Boukreev)

8. SYNCHRONOUS COMMUNICATIONS (MOOS, MUDS, IRCS)

`Fanderclai, Terri. Online interview. 11 Nov. 1996. LinguaMOO. 11 Nov. 1996 <telnet:// purple-crayon.media.mit.edu_8888>.`

In-Text Citation: `(Fanderclai)`

PART FOUR: A SAMPLE PAPER IN MLA STYLE

"I coulda been a contender," said Marlon Brando to Rod Steiger in the classic film *On the Waterfront.* According to film historian Steve Vineberg, this is "the most celebrated two-character exchange in the history of American movies." Both actors are graduates of Lee Strasberg's Actor's Studio, where for fifty years "the method" has been promoted as a technique for bringing powerful emotion to a performance.

In the research essay that follows, "Crying Real Tears: The History and Psychology of Method Acting,"* Christina Kerby starts by asking what many of us have wondered at one time or another: How do actors cry on command? For Christina, this is a personal question. As a young actor, she struggles to cry real tears. The key, she discovers, at least according to the method, may be to think long and hard about her dead cat.

What I love about Christina's research essay is not only how it demonstrates her genuine motive to find out something but also how her search takes her to the library, her acting class, interviews with acting teachers, the Internet, and her own past. All four sources of information—personal experience, reading, observation, and interview—are combined here. The result is a richly informative essay that goes way beyond a simple report on method acting. Christina describes how the technique revolutionized American theater in some particular ways. The writing is lively. The topic is fascinating. And the information is focused on Christina's research question: *How does method acting help actors* seem *to act with genuine feeling?*

*"Crying Real Tears: The History and Psychology of Method Acting" is reprinted with permission of Christina B. Kerby.

No title page is necessary. Double-space everything; use 1" margins all around. Pagination begins on the second page.

Christina Kerby

Professor Edelson

English 1A

30 September 1996

Crying Real Tears: The History and Psychology of Method Acting

"Acting is the most human of the arts. It's also the strangest, the weirdest."
Lee Strasberg (Hirsch 131)

Christina's lead nicely frames her purpose and suggests her motive for choosing the topic.

I have always envied people who can cry on command. The idea of letting yourself go, of throwing away all reservations and bawling your eyes out, kicking and screaming for no reason at all really appeals to me. I've tried it before. I was in a somewhat tragic scene a few years ago and thought it would be <u>so</u> dramatic if I could muster up a few big, glistening Bambi tears. I'd speak with a faltering dignity and I wouldn't dab at the corners of my eyes with a handkerchief because I'd <u>want</u> the audience to see my tears. I worked <u>hard</u> for my tears.

The tears never came. I got onstage and thought of my dead cat and unrequited love--things that really strike an emotional chord with me. But I was as dry as British pop humor. The fact that I <u>couldn't</u> cry made me want to cry. How do actors do it? The ones who

Kerby 2

sob and moan, "WHY!!? WHY!!?" or can lapse, in a second's time, from unbridled rage to howling laughter? How do they ride an emotional roller coaster seven days a week (and twice on Saturdays) for an ever-eager audience? Was it even possible for me to attain this incredible psychological feat? I resolved, then and there, to carry onions in my pockets for the remainder of the performances.

Christina has followed MLA style, which discourages hyphenating words at the ends of lines. Only words that already contain hyphens should be broken across lines.

As an aspiring actress (I use the term loosely), I am constantly searching for new ways to deal with roles, to get inside of a character's head. I've tried a variety of personal techniques, from "mood music" before a show to superstitions and good-luck charms to wearing a certain perfume to every performance. It wasn't until recently that I began to develop a curiosity about <u>method acting</u>. A few weeks ago, my acting teacher showed us a video about Lee Strasberg and the Actor's Studio. It was all very interesting: the advent of a revolution in American theater, new techniques and styles of acting and directing, all centered around some elusive concept that the narrator referred to as "the method." This could be the key to real tears!

Notice how Christina's first 4 paragraphs funnel down to her thesis.

Kerby 3

Webster's defines method acting as "a dramatic technique by which an actor seeks to gain complete identification with the inner personality of the character being portrayed. . . ." A weird connection. A psychological understanding. While this controversial approach has become one of the most hotly debated issues ever to cross the American stage, method acting has also played an immense role in shaping American cinema and theater, serving to re-define and in some ways, create, the role of the human psyche in acting.

The thesis is less argumentative than descriptive. It describes *a particular feature of method acting: its role in use of the psyche.*

The ideas behind method acting originated in Russia in the early twentieth century (Edwards 27). Constantin Stanislavski, a master of theater and acting technique, was one of the first to recognize the direct link between acting and the subconscious. Much as a Buddhist viewed enlightenment or nirvana, Stanislavski saw acting as a separate state of consciousness, a "paradise of art" (Edwards 271). He invented the emotional memory technique, in which an actor would recall past emotional memories or experiences as a means of recreating specific moods or emotions for a role (Edwards 270-272). Stanislavski's early ideas centered around finding a way to tap

Kerby 4

into the psyche in order to draw upon treasures that lie buried in the subconscious. This, he believed, was the key that would unlock the door to a great creative mood, a quest which he labeled "the superconscious through the conscious" (Hirsch 37).

As Stanislavski experimented with improvisation, however, he opted for a technique that would give the director more control over characterization and emotions. He shifted his emphasis from the internal to the external approach (Claudio), completely recentering his teachings around the method of physical actions, in which the actor shows the audience what she is feeling by what she does. The actor draws from her surroundings and imagination to create this emotional portrayal.

Information cited from an interview with Claudio (see also "Works Cited" entry).

Stanislavski recognized that certain human functions, such as heart rate and emotional reactions, are uncontrollable (Moore 11-12). A person will cry or laugh as a reaction to certain stimuli, such as a horrible memory or a funny joke. It is difficult to cry or even laugh on command unless there are stimuli to draw from. Therefore, the actor must construct emotions through actions, drawing from the set, the

The running head includes your last name and page number (without the abbreviation p.*), flush right.*

script, and the character. Stanislavski's system, or method of physical actions, stressed that actors should note how they react physically to different emotions, from doubling over when they are in pain to the extra little bounce in their step when they are happy. Stanislavski recognized that many of these actions are universal. As actor and director Buck Busfield pointed out, an actor can beg and plead in a thousand different languages, but the one way for an actor to really show the audience how desperate he is is to get down on his knees.

In the 1940s, American actor and director Lee Strasberg was beginning to explore techniques very similar to Stanislavski's. Strasberg, however, emphasized improvisation; he based his teachings almost purely around Stanislavski's early ideas regarding emotional recall, or affective memory (Hirsch 75). While Stanislavski stressed a physical portrayal, one step at a time, Strasberg taught actors to leap directly into a role by drawing on past emotional experiences, similar to accessing a file stored on a computer. He encouraged actors "not to act but to be themselves, to respond or react" ("Method").

Information cited from an Internet document (see also "Works Cited" entry).

Kerby 6

When American actress Stella Adler studied under Stanislavski in Paris in 1934, she returned to the United States with a wealth of new ideas about Stanislavski's revised teachings and his method of physical actions (Adler 117-122). "Stanislavski said we're doing it wrong" (Hirsch 79), Adler announced. Strasberg wasn't pleased at having his ideas challenged. The ensuing controversy marked the beginning of one of the greatest debates to face American theater.

As a result, many prominent American actors and directors were divided over Stanislavski's external approach versus Strasberg's focus on the internal at a raw, emotional level. In an attempt to completely separate his teachings from Stanislavski's system, Strasberg coined the term <u>method acting</u> (Garfield 168), which quickly became Strasberg's Americanized version of Stanislavski's earliest (and abandoned) teachings, <u>never</u> to be confused or associated with the <u>Stanislavski system</u> or <u>method of physical actions</u> (Edwards 261-262).

Strasberg went on to found the Actor's Studio, which attracted the most influential names in theater, among them Harold Clurman and Elia Kazan. Later, the studio would

Kerby 7

attract and educate Marlon Brando, James Dean, Marilyn Monroe, Sally Field, Shelly Winters, Matthew Broderick, and Paul Newman, among countless others (Hirsch). The studio was largely responsible for the growth and teaching of method in the acting community.

For a theater arts class, I was assigned a piece from Neil Simon's Brighton Beach Memoirs. I had a terrible cold and my cat had died the previous day, but I somehow managed to drag myself to class because I'd worked so hard to memorize that darn monologue, and under no circumstances was I going to pass up my chance to be dramatic. I got up in front of the whole class, the lights dimmed, and played the part of Nora as best as I knew how. And I thought I'd done a pretty good job. I was gathering up my props, expecting praise, when I heard my teacher say, "I want to see you do it again." She instructed me to think of how I feel when I am sad, and she gave me a human prop to interact with.

Is Christina's shift here to personal anecdote effective?

I thought of my dead cat and began the monologue again. And again. My teacher must have interrupted and made me start over eight times before I was able to get halfway through the piece, and by the time I did, I was completely distraught over the loss of my cat

Kerby 8

and the frustration and humiliation I felt at having to repeat the monologue. I stumbled through the piece and just hoped that I wasn't making a fool of myself and that I wouldn't have to do it again. When I finished, the class just sat there in silence. Then the applause started.

Did my classmates get some kind of sick pleasure out of watching me suffer? I cried all the way home. I knew that I had performed well but at what cost to my personal mental health? It wasn't until later that I realized that what I had experienced that day was method acting. Pretty scary stuff.

Is the transition here from personal narrative back to exposition handled well?

Strasberg's affective memory technique readily became the method's most widely disputed idea. Strasberg practiced what Stanislavski had tried to avoid by revising his teachings: he drove many actors and actresses to the point of near mental breakdown from the process of dredging up a painful memory over and over again. Sally Field recalled that Lee Strasberg had once forced her to remember an abusive relationship with her father, driving her to near hysteria. Finally, Strasberg told her that her performance was perfect, but then added, "Let's try that scene again."

Kerby 9

"Lee [Strasberg] crippled a lot of people. . . . Acting should be joyous, it should be pleasurable and easy if you do it right; but it wasn't with Strasberg," recalled Phoebe Brand, one of Strasberg's pupils (qtd. in Hirsch 77). It is apparent that the emotional and psychological toll from affective memory recall may outweigh the actual effectiveness of the technique; the actor risks becoming so immersed in his own complexities that he completely overlooks those of the character he is trying to portray.

During a phone interview, I asked my acting teacher, Ed Claudio, who studied under Stella Adler, whether he agreed with the ideas behind method acting. I could almost see him wrinkle his nose at the other end of the line. He described method as "self-indulgent," insisting that it encourages "island acting." Because of emotional recall, acting became a far more personal, internal art, and the actor began to move away from the script, often hiding the author's and <u>character's</u> purpose and intentions under her own. An actor who cannot control the outcome of affective memory on stage will, of course, take focus away from the script and from the other characters and

What are the advantages/disadvantages of Christina's inserting herself into her report of information from an interview?

place it all on herself. Audiences wish to be entertained and engaged but not necessarily knocked over by the gale-force winds of an actor's emotional indulgence.

For this reason, method is most effective during rehearsals, when an actor can try many different techniques until she masters, or at least understands, the emotions specific to her character. She can indulge in the intensity of an affective memory, study how the emotions affect her, and recreate the experience before an audience using Stanislavski's method of physical actions. This way, an actor is more likely to create a character from scratch, complete with a set of emotions unique to that character and more consistent with the author's intentions (if she is a good actor). The actor trains herself to react to an emotional stimuli, not only as she would personally but as the character would react.

In his series of lectures Method or Madness? Robert Lewis, a director and one of the method's leading critics, pointed out that method acting ". . . can lead to playing 'yourself' at the expense of the character rather than searching for the character in yourself" (79). "One senses," wrote Lewis, "that the actor is really feeling something, but that the emotion is a personal one, not necessarily related to the character or the particular play" (qtd. in Edwards 261). For example, an author may have created a character who grieves over the death of his father, while the actor playing the role may draw from affective memories of a painful divorce from his wife. Yes, the actor is feeling grief, but it is not the grief that the author had intended the character to feel. Lewis posed the question "Are . . . [actors] really living the part, or are they living themselves and adding the author's words to that life" (qtd. in Edwards 264)?

The use of qtd. in *signals that the Lewis quote is from the Edwards book; the original quote appeared elsewhere first.*

Method also tends to limit the actor's, and therefore the character's, range of emotions. Drawing purely from one's own emotions closes doors on a vast range of possibilities. For example, an actor who has never known immense frustration or rage would have a very difficult time portraying an explosive character while using the method.

While method acting had many drawbacks, it hit the acting community of the forties and fifties like a storm. Method, with its raw, emotional intensity, gave birth to a generation of actors. Because the method style

Kerby 12

of acting was unlike anything ever seen before, it was criticized as unrefined and even dangerous. James Dean did not merely *play* the role of a teenage rebel, he *was* a teenage rebel. Marilyn Monroe carried a level of sexual power to the screen that brought fire to the hearts of teenage America and fear to the hearts of parents! One author attributed method acting to the Beat Generation rebellion, quipping, "They have seen no signs of a search for spiritual values in a generation . . . whose interests have ranged all the way from bebop to rock and roll, from hipsterism to Zen Buddhism, from vision-inducing drugs to method acting" (Holmes 16).

Perhaps method acting was considered dangerous because actors and actresses were experiencing a new height of emotional awareness that they hadn't thought possible. Up until this point, acting technique had centered almost entirely around action and physicality. The earliest tragedians hid their faces behind huge masks. More modern actors were typically selected based on physical stature or voice quality, *not* emoting ability. Described as a "Renaissance in acting" (Easty 4), method not only made accessible an array of new techniques, it broadened the range of possibilities that an actor could explore, therefore expanding the art of acting itself. Perhaps the new and unexplained are what frightened many traditionalists away from the method.

For all of the method's weaknesses, there must have been some strange allure that kept audiences glued, actors immersed, and Lee Strasberg obsessed. It offered the exploration of a new type of character, complete with rough-hewn edges and cutting emotional intensity. Success stories like Marilyn Monroe, James Dean, and Marlon Brando, who were revered for their fresh, spontaneous style of acting, spurred a revolution in cinema when acting suddenly became more natural and more accessible. James Dean's ". . . control switch was permanently on 'off'; his complexes were always leaking through the holes in his delicate sensitivities," making him captivating and alluring to an audience of teenage "rebels" (Vineberg 188).

This is a great interview quote—short, memorable, well said.

Dave Pierini, a local Sacramento actor, pointed out, "You can be a good actor without using the method, but you cannot be a good actor without at least understanding it." Actors are perhaps some of the greatest scholars of the human psyche because they

Kerby 14

devote their lives to its study. Aspiring artists are told to "get inside of the character's head." They are asked, "How would the character feel? How would the character <u>react</u>?" To those who claim that method acting is out of date or ineffective, I argue that to understand the complexity of a character's mind, an actor must first understand himself. Perhaps this is the point where the method is most useful to actors. No matter what system he is using, an actor must understand emotions before he can begin to recreate them.

An actor is also forced to get to know himself better, and in many different ways, with each character he portrays. He may find a little bit of himself in the paranoid schizophrenic, the homicidal maniac, or the forlorn lover. And, of course, every actor will bring a little bit of himself to a role. Whether he uses the method or not will very likely affect just how much of himself he brings.

While method may have helped to ensure the livelihood of Hollywood "shrinks," it has also opened doors that actors never knew existed. Acting is no longer purely about poise, voice quality, and diction. It is also about feeling the part, about understanding

Kerby 15

the emotions that go into playing that part, and about bringing those emotions to life within the character. Whether an actor uses Stanislavski's method of physical actions to unlock the door to her subconscious or whether she attempts to stir up emotions from deep within herself using Strasberg's method, the actor's goal is to create a portrayal that is truthful. It's possible to pick out a bad actor from a mile away: one who does not understand the role because she does not understand the emotions necessary to create it. Or perhaps she simply lacks a means of tapping into them.

If genuine emotion is what the masses want, method acting may be just what every starstruck actress needs. Real tears? No problem. The audience will never know that Juliet was not lamenting the loss of her true love, Romeo, but invoking the memory of her favorite cat, Fluffy, who died tragically in her arms.

Christina's ending defies the usual convention of summarizing the main points. What makes it more effective?

Kerby 16

Works Cited

Adler, Stella. The Technique of Acting. New York: Bantam, 1988.

Busfield, Buck. Personal interview. 13 September 1996.

Claudio, Ed. Telephone interview. 9 December 1995.

---. "The History of Method Acting and Modern Theater." Letter. Sacramento Bee. 4 June 1995, sec. 4: 2.

Easty, Edward Dwight. On Method Acting. Orlando, FL: House of Collectibles, 1978.

Edwards, Christine. The Stanislavski Heritage. New York: New York UP, 1965.

Garfield, David. The Actor's Studio: A Player's Place. New York: Macmillan, 1980.

Hirsch, Foster. A Method to Their Madness. New York: Norton, 1984.

Holmes, John Clellon. "The Philosophy of the Beat Generation." The Beats. Ed. Seymour Krimm. Greenwich, CT: Fawcett, 1960.

Lewis, Robert. Method or Madness? New York: Samuel French, 1958.

"Method Acting." Palace: The Classic Movie Site. 19 June 1996 <http://www.scruz.net/~mmills/palace/method.html>

Moore, Sonia. The Stanislavski System. New York: Viking, 1974.

Pierini, Dave. Personal interview. 2 January 1996.

Vineberg, Steve. Method Actors. New York: Schirmer, 1991.

Double-space the "Works Cited" page; indent the second line of each entry 5 spaces.

Like many online sources, this one lacks some information; cite as much as you have.

APPENDIX B

□ ■ □

Guide to APA Style

The Modern Language Association (MLA) author/page number system for citing borrowed material, described in Appendix A, is the standard for most papers written in the humanities, though some disciplines in the fine arts as well as history and philosophy may still use the footnote system. Confirm with your instructor that the MLA system is the one to use for your paper.

Another popular documentation style is the American Psychological Association (APA) author/date system. APA style is the standard for papers in the social sciences as well as biology, earth science, education, and business. In those disciplines, the currency of the material cited is often important.

I think you'll find APA style easy to use, especially if you've had some practice with MLA. Converting from one style to the other is easy. Basically, the APA author/date style cites the author of the borrowed material and the year it appeared. A more complete citation is listed in the "References" (the APA version of MLA's "Works Cited") at the back of the paper. (See the sample APA-style paper in Part Four of this appendix.)

The *Publication Manual of the American Psychological Association** is the authoritative reference on APA style. Though the information in the section that follows should answer your questions, check the manual when in doubt. A handy "cribsheet," which provides highlights from the *Publication Manual,* is also available online at the following Web address:

http://www.gasou.edu/psychweb/tipsheet/apacrib.htm

**Publication Manual of the American Psychological Association,* 4th ed. (Washington, DC: APA, 1994).

PART ONE:
HOW THE ESSAY SHOULD LOOK

Page Format. Papers should be double spaced, with at least 1" margins on all sides. As always, use a fresh ribbon or cartridge on your typewriter or printer and avoid fancy typefaces. Number all pages consecutively, beginning with the title page; put the page number in the upper-righthand corner. Above the page number, place an abbreviated title of the paper on every page, in case pages get separated. As a rule, all paragraphs of text should be indented five spaces.

Title Page. Unlike a paper in MLA style, an APA-style paper often has a separate title page, containing the following information: the title of the paper, the author, and the author's affiliation (what university she is from). At the bottom of the title page, in uppercase letters, you may also include a *running head,* or an abbreviation of the title (fifty characters or less, including spaces). Repeat this shortened title at the top of each manuscript page, along with the page number. Each line of information should be centered and double spaced.

Abstract. Though it's not always required, many APA-style papers include a short abstract (no longer than 120 words) following the title page. An abstract is essentially a short summary of the paper's contents, usually in about 100 words. This is a key feature, since it's usually the first thing a reader encounters. The abstract should include statements about what problem or question the paper examines and what approach it follows; the abstract should also cite the thesis and significant findings. Type the title "Abstract" at the top of the page. Type the abstract text in a single block, without indenting.

Body of the Paper. The body of the paper begins with the centered title, followed by a double-space and then the text. A page number (usually an abbreviated title and *3* if the paper has a title page and abstract) should appear in the upper-righthand corner.

You may find that you want to use headings within your paper. If your paper is fairly formal, some headings might be prescribed, such as "Introduction," "Method," "Results," and "Discussion." Or create your own heads to clarify the organization of your paper.

If you use headings, the APA recommends a hierarchy like this:

Centered Upper- and Lowercase

<u>Centered, Underlined, Upper- and Lowercase</u>

<u>Flush Left, Underlined, Upper- and Lowercase</u>

<u>Indented, underlined, lowercase; ends with period.</u>

Note that none of these headings is typed in **bold** or *italic* type; rather, underlining is used to distinguish the second-, third-, and fourth-level headings. This is the format recommended by APA style. So even if your computer or word processor can produce different kinds of type, you should stick with underlining. (This is also true for titles in your "References" section.) Check with your instructor to make sure of any preferences she may have.

References Page. All sources cited in the body of the paper are listed alphabetically by author (or title, if anonymous) on the page titled "References." This list should begin a new page. Each entry is double spaced; begin each first line flush left, and indent subsequent lines five to seven spaces. Explanation of how to cite various sources in the references follows (see "Part Three: Preparing the 'References' List").

Appendix. This is a seldom used feature of an APA-style paper, though you might find it helpful for presenting specific or tangential material that isn't central to the discussion in the body of your paper: a detailed description of a device described in the paper, a copy of a blank survey, or the like. Each item should begin on a separate page and be labeled "Appendix" followed by "A," "B," and so on, consecutively.

Notes. Several kinds of notes might be included in a paper. The most common is *content notes,* or brief commentaries by the writer keyed to superscript numbers in the body of the text. These notes are useful for discussion of key points that are relevant but might be distracting if explored in the text of your paper. Present all notes, numbered consecutively, on a page titled "Footnotes." Each note should be double spaced. Begin each note with the appropriate superscript number, keyed to the text. Indent each first line five to seven spaces; consecutive lines run the full page measure.

Tables and Figures. The final section of an APA-style paper features tables and figures mentioned in the text. Tables should all be double spaced. Type a table number at the top of the page, flush left. Number tables "Table 1," "Table 2," and so on, corresponding to the order they are mentioned in the text. A table may also include a title. Each table should begin on a separate page.

Figures (illustrations, graphs, charts, photographs, drawings) are handled similarly to tables. Each should be titled "Figure" and numbered consecutively. Captions may be included, but all should be typed on a separate page, clearly labeled "Figure Captions," and listed in order. For example:

Figure Captions

Figure 1: A photograph taken in the 1930s by Dorthea Lange.

Figure 2: Edward Weston took a series of green pepper photographs like this. This is titled "No. 35."

PART TWO: CITING SOURCES IN YOUR ESSAY

1. WHEN THE AUTHOR IS MENTIONED IN THE TEXT

The author/date system is pretty uncomplicated. If you mention the name of the author in text, simply place the year her work was published in parentheses immediately after her name. For example:

Herrick (1992) argued that college testing was biased against minorities.

2. WHEN THE AUTHOR ISN'T MENTIONED IN THE TEXT

If you don't mention the author's name in the text, then include that information parenthetically. For example:

A New Hampshire political scientist (Bloom, 1992) recently studied the state's presidential primary.

Note that the author's name and the year of her work are separated by a comma.

3. WHEN TO CITE PAGE NUMBERS

If the information you're citing came from specific pages (or chapters or sections) of a source, that information may also be included in the parenthetical citation. Including page numbers is essential when quoting a source. For example:

The first stage of language acquisition is called "caretaker speech" (Moskowitz, 1985, pp. 50-51), in which children model their parents' language.

The same passage might also be cited this way if the authority's name is mentioned in the text:

```
Moskowitz (1985) observed that the first stage of
language acquisition is called "caretaker speech"
(pp. 50-51), in which children model their
parents' language.
```

4. HOW TO CITE TWO OR MORE AUTHORS

When a work has two authors, always mention them both whenever you cite their work in your paper. For example:

```
Allen and Oliver (1982) observed many cases of
child abuse and concluded that maltreatment
inhibited language development.
```

If a source has more than two authors but less than six, mention them all the first time you refer to their work. However, any subsequent references can include the surname of the first author followed by the abbreviation *et al.* When citing works with more than six authors, *always* use the first author's surname and *et al.*

5. HOW TO CITE A WORK WITH NO AUTHOR

When a work has no author, cite an abbreviated title and the year. Place article or chapter titles in quotation marks, and underline book titles. For example:

```
The editorial ("Sinking," 1992) concluded that the
EPA was mired in bureaucratic muck.
```

6. HOW TO CITE TWO OR MORE WORKS BY THE SAME AUTHOR

Works by the same author are usually distinguished by the date; works are rarely published the same year. But if they are, distinguish among works by adding an *a* or *b* immediately following the year in the parenthetical citation. The reference list will also have these suffixes. For example:

```
Douglas's studies (1986a) on the mating habits of
lobsters revealed that the females are dominant.
He also found that the female lobsters have the
uncanny ability to smell a loser (1986b).
```

This citation alerts readers that the information came from two studies by Douglas, both published in 1986.

7. CITING AN INSTITUTIONAL AUTHOR

When citing a corporation or agency as a source, simply list the year of the study in parentheses if you mention the institution in the text:

> The Environmental Protection Agency (1992) issued an alarming report on ozone pollution.

If you don't mention the institutional source in the text, spell it out in its entirety, along with the year. In subsequent parenthetical citations, abbreviate the name. For example:

> A study (Environmental Protection Agency [EPA], 1992) predicted dire consequences from continued ozone depletion.

And later:

> Continued ozone depletion may result in widespread skin cancers (EPA, 1992).

8. CITING INTERVIEWS, E-MAIL, AND LETTERS

Interviews and other personal communications are not listed in the references at the back of the paper, since they are not *recoverable data,* but they are parenthetically cited in the text. Provide the initials and surname of the subject (if not mentioned in the text), the nature of the communication, and the complete date.

> Nancy Diamonti (personal interview, November 12, 1990) disagrees with the critics of Sesame Street.

9. NEW EDITIONS OF OLD WORKS

For reprints of older works, include both the year of the original publication and that of the reprint edition (or the translation).

> Pragmatism as a philosophy sought connection between scientific study and real people's lives (James 1906/1978).

PART THREE: PREPARING THE "REFERENCES" LIST

All parenthetical citations in the body of the paper correspond to a complete listing of sources on the "References" page. The format for this section was described earlier in this appendix (see "References Page").

1. ORDER OF SOURCES

List the references alphabetically by author or by the first key word of the title if there is no author. The only complication may be if you have several articles or books by the same author. If the sources weren't published in the same year, list them in chronological order, the earliest first. If the sources were published in the *same* year, include a lowercase letter to distinguish them. For example:

```
Lane, Barry. (1991a). Verbal medicine . . .
Lane, Barry. (1991b). Writing . . .
```

2. ORDER OF INFORMATION

A reference to a periodical or book in APA style includes this information, in order: author, date of publication, article title, periodical title, and publication information.

Author. List *all* authors—last name, comma, and then first name along with an initial, if any. Use commas to separate authors' names; add an ampersand (&) before the last author's name. When citing an edited book, list the editor(s) in place of the author, and add the abbreviation *Ed.* or *Eds.* in parentheses following the last name. End the list of names with a period.

Date. List the year the work was published, along with the date if it's a magazine or newspaper (see "Sample References," following), in parentheses, immediately after the last author's name. Add a period after the closing parenthesis.

Article or Book Title. APA style departs from MLA, at least with respect to periodicals. In APA style, only the first word of the article title is capitalized, and it is not underlined or quoted. Book titles, on the other hand, are underlined; capitalize only the first word of the title and any subtitle. End all titles with periods.

Periodical Title and Publication Information. Underline the complete periodical title; type it using both uppercase and lowercase letters. Add the volume number (if any), also underlined. Separate the title and volume number with a comma (e.g., *Journal of Mass Communication, 10,* 138–150). If each issue of the periodical starts with page 1, then also include the issue number in parentheses immediately after the volume number (see examples following). End the entry with the page numbers of the article. Use the abbreviation *p.* or *pp.* if you're citing a newspaper or magazine; omit it if you're citing a journal.

For books, list the city of publication (adding the state or country if the city is unfamiliar; use postal abbreviations) and the name of the publisher; separate the city and publisher with a colon. End the citation with a period.

Remember that the first line of each citation should begin flush left and all subsequent lines should be indented five to seven spaces. Double-space all entries.

Sample References

1. A JOURNAL ARTICLE

Cite a journal article like this:

Blager, Florence B. (1979). The effect of intervention on the speech and language of children. Child Abuse and Neglect, 5, 91-96.

In-Text Citations: (Blager, 1979)

If the author is mentioned in the text, just parenthetically cite the year: Blager (1979) stated that . . .

If the author is quoted, include the page number(s): (Blager, 1979, p. 992)

2. A JOURNAL ARTICLE NOT PAGINATED CONTINUOUSLY

Most journals begin on page 1 with the first issue of the year and continue paginating consecutively for subsequent issues. A few journals, however, start on page 1 with each issue. For these, include the issue number in parentheses following the volume number:

Williams, John, Post, Albert T., & Stunk, Fredrick. (1991). The rhetoric of inequality. Attwanata, 12(3), 54-67.

In-Text Citations: (Williams, Post, & Stunk, 1991)

Subsequent citations would use *et al.:* (Williams et al., 1991)

If quoting material, include the page number(s): (Williams et al., 1991, pp. 55-60)

3. A MAGAZINE ARTICLE

Maya, Pines. (1981, December). The civilizing of Genie. Psychology Today, pp. 28-34.

In-Text Citations: (Maya, 1981)

Maya (1981) observed that . . .

If quoting, include the page number(s): (Maya, 1981, p. 28)

4. A NEWSPAPER ARTICLE

Honan, William. (1991, January 24). The war affects Broadway. New York Times, pp. C15-16.

In-Text Citations: (Honan, 1991)

Honan (1991) argued that . . .

Honan (1991) said that "Broadway is a battleground" (p. C15).

5. A BOOK

Lukas, Anthony J. (1986). Common ground: A turbulent decade in the lives of three American families. New York: Random House.

In-Text Citations: (Lukas, 1986)

According to Lukas (1986), . . .

If quoting, include the page number(s).

6. A BOOK OR ARTICLE WITH MORE THAN ONE AUTHOR

Rosenbaum, Alan, & O'Leary, Daniel. (1978). Children: The unintended victims of marital violence. American Journal of Orthopsychiatry, 4, 692-699.

In-Text Citations: (Rosenbaum & O'Leary, 1978)

Rosenbaum and O'Leary (1978) believed that . . .

If quoting, include the page number(s).

7. A BOOK OR ARTICLE WITH AN UNKNOWN AUTHOR

New Hampshire loud and clear. (1992, February 19). The Boston Globe, p. 22.

In-Text Citations: ("New Hampshire," 1992)

Or mention the source in text: In the article "New Hampshire loud and clear" (1992), . . .

If quoting, provide the page number(s), as well.

A Manual of Style (16th ed.). (1993). Chicago: University of Chicago Press.

In-Text Citations: (Manual of Style, 1993)

According to the Chicago Manual of Style (1993), . . .

If quoting, include the page number(s).

8. A BOOK WITH A CORPORATE AUTHOR

American Red Cross. (1979). Advanced first aid and emergency care. New York: Doubleday.

In-Text Citations: (Advanced First Aid, 1979)

The book Advanced First Aid and Emergency Care (1979) stated that . . .

If quoting, include the page number(s).

9. A BOOK WITH AN EDITOR

Crane, R. S. (Ed.). (1952). Critics and criticism. Chicago: University of Chicago Press.

In-Text Citations: (Crane, 1952)

In his preface, Crane (1952) observed that . . .

If quoting, include the page number(s).

10. A SELECTION IN A BOOK WITH AN EDITOR

McKoen, Richard. (1952). Rhetoric in the Middle Ages. In R. S. Crane (Ed.), Critics and criticism (pp. 260-289). Chicago: University of Chicago Press.

In-Text Citations: (McKoen, 1952)
McKoen (1952) argued that . . .
If quoting, include the page number(s).

11. A REPUBLISHED WORK

James, William. (1978). Pragmatism. Cambridge, MA: Harvard University Press. (Original work published 1907).

In-Text Citations: (James 1907/1978)
According to William James (1907/1978), . . .
If quoting, include the page number(s).

12. A BOOK REVIEW

Dentan, R. K. (1989). A new look at the brain [Review of The dreaming brain]. Psychiatric Journal, 13, 51.

In-Text Citations: (Dentan, 1989)
Dentan (1989) argued that . . .
If quoting, include the page number(s).

13. A GOVERNMENT DOCUMENT

U.S. Bureau of the Census. (1991). Statistical abstract of the United States (111th ed.). Washington, DC: U.S. Government Printing Office.

In-Text Citations: (U.S. Bureau, 1991)
According to the U.S. Census Bureau (1991), . . .
If quoting, include the page number(s).

14. A LETTER TO THE EDITOR

Hill, Anthony C. (1992, February 19). A flawed history of blacks in Boston [Letter to the editor]. The Boston Globe, p. 22.

In-Text Citations: (Hill, 1992)

Hill (1992) complained that . . .

If quoting, include page number(s).

15. A PUBLISHED INTERVIEW

Personal interviews are usually not cited in an APA-style paper, unlike published interviews.

Cotton, Peter (1982, April). [Interview with Jake Tule, psychic]. Chronicles Magazine, pp. 24-28.

In-Text Citations: (Cotton, 1982)

Cotton (1982) noted that . . .

If quoting, include the page number(s).

16. A FILM OR VIDEOTAPE

Hitchcock, A. (Producer and Director). (1954). Rear window [Film]. Los Angeles, CA: MGM.

In-Text Citations: (Hitchcock, 1954)

In Rear Window, Hitchcock (1954) . . .

17. A TELEVISION PROGRAM

Burns, K. (Executive Producer). (1996). The West. New York and Washington, DC: Public Broadcasting Service.

In-Text Citations: (Burns, 1996)

In Ken Burns's (1996) film, . . .

For an episode of a television series, use the scriptwriter as the author, and provide the director's name after the program title. List the producer's name last.

Duncan, D. (1996). Episode two (S. Ives, Director). In K. Burns (Producer), The West. New York and Washington, DC: Public Broadcasting Service.

In-Text Citations: (Duncan, 1996)
In the second episode, Duncan (1996) explores . . .

18. A MUSICAL RECORDING

Wolf, K. (1986). Muddy roads. On Gold in California [CD]. Santa Monica, CA: Rhino Records.

In-Text Citations: (Wolf, 1986)
In Wolf's (1986) song, . . .

19. A COMPUTER PROGRAM

TLP.EXE Version 1.0 [Computer software]. (1991). Hollis, NH: Transparent Language.

In-Text Citations: (TLP.EXE Version 1.0, 1991)
In TLP.EXE (1991), a pop-up window . . .

20. ONLINE SOURCES

The APA, like the MLA, is currently grappling with the rapidly changing nature of electronic sources. As a result, citation methods continue to evolve. To keep up with some of these changes, I recommend that you check the following Web sites, each of which have links to documents and proposals on both APA and MLA formats:

http://www.cc.emory.edu/WHSCL/citation.formats.html

http://alpha.cmns.mnegri.it/WWW_HTML_IT/extern_resources/citing_internet_ref.html

According to the latest edition of APA's *Publication Manual,* the basic citation format for online sources doesn't differ too much from that for print sources, except for the addition of information about how the material can be retrieved from the Internet. Each citation begins with the *author's name* (if any); the *title* of the article or work; a designation in brackets of *the type of source,* usually [Online]; the *length,* if it's an article (*pars.* or *pgs.*); and the *protocol* (http, FTP, Telnet, etc.) and *pathway* to retrieve the document or the URL if it's a Web-based source. End the citation by including the *access date* in brackets. For example:

An Article

Adler, J. (1996). Save endangered species, not the endangered species act. Intellectual Ammunition [Online], 2 pgs. Available HTTP: http://www.heartland.org/05jnfb96.htm [1996, October 12]

In-Text Citations: (Adler, 1996)

According to Adler (1996), . . .

If quoting, include the page number(s).

An Electronic Text

Encyclopedia Mythica. (1996). [Online]. Available HTTP: http://www.pantheon.org/myth [1996, Oct. 11]

In-Text Citations: (Encyclopedia Mythica, 1996)

The Encyclopedia Mythica (1996) presents . . .

If the text is an electronic version of a book published in print earlier, include the original publication date in parentheses following the title: (Orig. pub. 1908)

A Part of a Work

Hunter, J. (No date). Achilles. In Encyclopedia Mythica, [Online]. Available HTTP: http://www.pantheon.org/myth/achill [1996, Oct. 11]

In-Text Citations: (Hunter, no date)

According to Hunter (no date), Achilles was . . .

If quoting, include the page or paragraph number(s), if any.

An Online Journal

Haynes, C., & Holmevik, J. (1996). Enhancing pedagogical reality with MOOs. Kairos: A Journal for Teachers of Writing in a Webbed Environment [Online], 1(2), 1 pg. Available HTTP: http://english.ttu.edu/kairos/1.2/index.html [1997, June 10]

In-Text Citations: (Haynes & Holmevik, 1996)

Haynes and Holmevik (1996) recently pointed out . . .

If quoting, include the page or paragraph number(s), if any.

Discussion Lists

Pryor, C. (1996, March 20). Referencing Internet material. Megabyte University Discussion List [Online]. Available E-mail: mbu-l@unicorn.acs.ttu.edu [1996, March 22]

In-Text Citations: (Pryor, 1996)

In a recent listserv discussion, Pryor (1996) . . .

21. CD-ROM ENCYCLOPEDIAS AND DATABASES

Psychotherapy. (1994). In Microsoft Encarta (1994 ed.), [CD-ROM]. Available: Microsoft Corp. [1996, June 15].

In-Text Citation: ("Psychotherapy," 1994)

Kolata, G. (1996). Research links writing style to the risk of Alzheimer's [CD-ROM]. New York Times, p. 1A. Abstract from: UMI-Proquest/Newspaper Abstracts [10 July 1997].

In-Text Citations: (Kolata, 1996)

Kolata (1996) believes that Alzheimer's . . .

If quoting, include the page or paragraph number(s), if any.

Haden, C. A. (1996). Talking about the past with preschool siblings [CD-ROM]. Dissertation Abstracts International, 56, 1324-1325. Abstract from: UMI-Proquest/Dissertation Abstracts Ondisc [12 October 1996].

In-Text Citations: (Haden, 1996)

Haden (1996) argues that . . .

If quoting, include the page or paragraph number(s), if any.

PART FOUR: A SAMPLE PAPER IN APA STYLE

The thing that struck me immediately about Carolyn Nelson's essay* on the Endangered Species Act is how she manages to *be* personal without *getting* personal. Though she never uses the first-person singular here and doesn't include any autobiographical material, Carolyn has a strong presence throughout the piece. That presence isn't simply registered by her clear point of view—she thinks the Endangered Species Act tramples on property rights, among other things. We also hear her voice through her control of quotation, her knack of finding her own way of saying things, her strong writing voice, and her subtle ability to keep up a conversation with both her sources and her readers. As you read Carolyn's essay, mark those passages in which her presence is felt most strongly. What is she doing there?

"The Endangered Species Act: Protecting Both Warblers and the Trotters of the World" is an argumentative essay, which makes it a conventional type of research paper. Carolyn does a fine job of making her case—so much so that I find I want to argue with her! Compare her paper with Christina Kerby's more exploratory research essay on method acting (see Appendix A). How does each piece establish a slightly different relationship with you as a reader? Does each writer seem to have a different relationship to her authorities and information? How would you describe the differences, if you notice any? What might explain them?

*"The Endangered Species Act: Protecting Both Warblers and the Trotters of the World" is reprinted with permission of Carolyn Nelson.

APA style usually requires a title page.

Endangered Species 1

Running Head: ENDANGERED SPECIES

The running head*—an abbreviation of the title—appears here and then on every page.*

The Endangered Species Act:

Protecting Both Warblers and the

Trotters of the World

Carolyn Nelson

Sacramento Community College

If your instructor requires it, include an abstract of less than 120 words. Double-space; don't indent the first line.

Abstract

The federal Endangered Species Act (ESA), which became law in 1973, significantly broadened the government's power to limit human activity that appears to threaten the recovery of listed species. Since its passage 23 years ago, only 6 species have fully recovered, and half of the listed species have formal "recovery plans" (Horton, 1992). The ESA has caused significant unemployment in certain impacted areas and, because it places equal value on all species listed, does not provide a means for balancing economic and ecological impacts. Under current rules, the ESA also does not justly compensate private property owners. Reform of the act is essential, or it will lose public support.

Endangered Species 3

Margins should be at least 1" all around.

Center and repeat the full title; double-space; then begin the text.

The Endangered Species Act:
Protecting Both the Warblers
and the Trotters of the World

Some may call David Trotter of Austin, Texas, lucky. Owning a significant amount of land on Canyon Creek, Trotter shares his property with a wide assortment of endangered animals: golden-checked warblers, Tooth Cave ground beetles, Kretschmarr Cave mold beetles, Tooth Cave pseudoscorpions, and Tooth Cave spiders. Because so many rare and wondrous creatures inhabit the land, many assume Trotter is grateful for the location of his property. After all, golden-checked warblers are scarce in Texas (Reiger, 1995).

So it may come as a surprise to learn that Trotter is far from thrilled. And it is for one reason: the Endangered Species Act. As a result of the endangered species' presence, Trotter has been required to "donate" 721 acres of his land to "ensure that it is protected in perpetuity" (Reiger, 1995, p. 17). Another 873 acres were demanded after being deemed "reasonable and prudent measures necessary and appropriate to minimize incidental take of warblers" (p. 17). The donated land serves as a refuge for the

Carolyn avoids an abstract or general opening; she makes her topic more compelling by showing its impact on one man.

warblers, and Trotter bears a hunk of the monetary burden for the sanctuary.

All this leads one to wonder: Is the Endangered Species Act a brilliant piece of legislation, from which every species will flourish, or, rather, a plan which is solely beneficial to endangered animals and their populations, completely indifferent to human activity? Should animals that in no way aid in the advancement of society, and whose extinction would in no way harm the ecosystem, take precedence over economic and industrial growth? And finally, is it reasonable for the United States government to seize property in the name of the act, without just compensation, in light of the Fifth Amendment?

Carolyn neatly raises the three questions her paper will answer and structures her paper around them.

Simply put, the answer is no.

Prior to the act's establishment, such questions were not so relevant. The main goal of the legislation was to protect popular endangered species, like the bald eagle. But with the passage of the act in 1973, the government broadened its environmental concerns. More detailed than any other major piece of endangered species legislation before it, the Endangered Species Act (ESA) established two major new precedents: the

Fish and Wildlife Service and the National Marine Fisheries Service are in charge of choosing the species truly threatened or endangered, and once the two agencies have listed a species as threatened or endangered, no branch of the federal government can proceed with a project that may harm a member of a listed species (Chadwick, 1995). According to John Fahey (1990), the ESA of 1973 is designed to "keep all forms of life, . . . from Bambi to pond slime, from going extinct" (p. 86).

Direct quotes require citing page numbers.

Although its purpose is to preserve all forms of life, the act's success rate does not live up to its promises. In the ESA's 23 years on the books, only 6 species have fully recovered (Carpenter, 1995). Thirty-six hundred more species await approval, and only one half of the listed species have "formal recovery plans" (Horton, 1992, p. 72). The act is clearly ineffective in achieving its goals.

The goal of the act, obviously rarely reached, is to revive species nearing extinction. But it does not follow that we should ignore the welfare of other species in the process. The ESA should be working to create an environment where humans and the bald eagle can live together. Right now, once a species makes the list, a critical habitat is declared, recovery plans are made, and any takings are outlawed. Sounds great. But one must look closer: A taking includes harassing, harming, pursuing, hunting, shooting, wounding, trapping, killing, capturing, or collecting (Lacy & Lepper, 1996a). Suddenly, a human can barely breathe around an endangered animal. The act does not provide an environment where humans and endangered species can coexist.

The lowercase a *signals that this is one of 2 or more articles by these authors and published in 1996 (see also the "References" entry).*

The laws set forth by the ESA do not support peaceful coexistence and result in a significant rise in unemployment rates for certain areas. Many people lose their jobs every year due to the constraints of the act. For example, according to Chadwick (1995), in response to a government report recommending the closure of 8 million acres of land from chain saws, loggers of the Pacific Northwest lost thousands of jobs and many business dollars. The unemployment rate in the region boomed, and loggers responded with protest, "Save a Logger, Eat an Owl" and "We're the endangered ones" (p. 26). Such situations are not uncommon, but proponents of the act still feel it is just.

Some may say that there's a certain cost for the survival of the endangered species, and if that cost is the job of some natives in Oregon, so be it. A senior staff member with a national environmental organization commented, "You can't go out and save every single person, every single community. It's not possible" ("Counter Movement," 1993, p. 11). Of course, the senior staff member would be right, if the act, as it stands now, was working. But it is not. With a poor track record, it makes little sense to compromise jobs and communities for the act. Furthermore, it makes little sense to support the continuing of the ESA without change. If it was weakened a little and took into account a human element, the act might be better armed to achieve its goal: the preservation of endangered and threatened species.

Carolyn uses strong quotes throughout her essay.

Not only does the act make a far-reaching attempt to save other species with disregard of human activity, but the ESA also makes an effort to save those species that make questionable contributions to the ecosystem. David Bjerklie (1992) wrote that "the hallmark of the act is that it considers species to be of equal 'incalculable value.'

Carolyn has followed APA style, which discourages hyphenating words at the ends of lines. Only words that already contain hyphens should be broken across lines.

Endangered Species 8

Consequently, a tiny button cactus merits the same protection accorded a glamorous wildlife-poster species like the bald eagle" (p. 20). This aspect of the legislation is one of great controversy. Some feel that it is important to assign each creature equal protection; however, this does not make sense. In a case lacking absolutes, the United States Supreme Court currently employs a policy of weighing costs and benefits in various areas. So <u>why</u> should there be an absolute in the case of endangered species?

Often enough, a small subspecies of questionable worth holds up a large project. When the act was first passed by Congress, a tiny fish called the snail darter held up a costly project: Telico Dam. A multimillion dollar project, the dam was already half complete when construction came to a halt. Because a cost-benefit system was not in use, the darters caused significant delays in production (Lepper & Lacy, 1996b). Such a small creature need not delay the advancement of society, especially when it could be removed from its habitat without harming the ecosystem.

Aside from the economic issues, not all species are of importance in their ecosystem. Daniel Simberloff, of Florida State University, stated that "there are species that could disappear without really impacting the ecosystem" (Carpenter, 1995, p. 44). Increasingly, scientists have questioned the act's premise that all species are created equal. Carpenter (1995) cited new studies that have argued that nature may build "spare parts" into its ecosystems (p. 45). With this new knowledge that all creatures are not necessary for the survival of an ecosystem, Congress should recognize the problem with the ESA's blanket protection policy.

In addition to protecting creatures that would not be missed, the act protects threatened subspecies of a species that, as a whole, might not be endangered. For example, Chadwick (1995) noted that there are 297 different kinds of freshwater mussels in the United States. Fifty-six of them are listed as threatened or endangered, 74 are declining, and 21 may already be extinct. Does each subspecies deserve as much protection as the bald eagle? Would their absence have as great an impact on the United States? No. It could be argued that each subspecies has the potential to be the next miracle cure for AIDS, cancer, or other diseases, but the miracle-cure theory is not worth the trouble. To chase after every single lead on a cure, to test every single species for a cure, is an unreasonable notion. It would be an impossible task. Society would benefit more from a greater freedom--to move how and where it pleases--than if it gave up the freedom for an unsound idea.

A clear violation of the United States Constitution, the ESA rarely compensates landowners for their losses due to the act. The Fifth Amendment states, "Nor shall property be taken for public use without just compensation" (Reiger, 1995, p. 16). Despite the explicitness of the constitutional violation, some still argue that compensation should not be awarded. An article in National Wildlife magazine went so far as to declare, "The whole idea that the government needs to pay landowners not to do bad things is ridiculous" (Reiger, 1995, p. 17). This argument is wrong. Nowhere in the Constitution does it say "Nor shall property be taken for public use without just compensation . . . unless it's a really, really good cause." A

Carolyn does not suppress arguments from sources she disagrees with but uses them to develop her own position.

project that might seem trivial to a member of the Save the Owls foundation could be of great significance to a particular landowner or community. Because good and bad are such relative concepts, the magazine's argument cannot be taken seriously.

Other objections are also unreasonable. John Echeverria, of the National Audubon Society, believes that Congress is wasting its time with revisions. "The pending takings bills are costly, wasteful, destructive pieces of legislation. The Constitution already protects private property rights. The Supreme Court has established standards for addressing Fifth Amendment takings claims. There's no need for legislation on this subject" (Jost, 1995, p. 515). But Michael Rubin rightly counters that "the bottom line is that courts very rarely hold that a regulation is a taking" (p. 516).

Congress is on a quest to make compensation a more reachable indemnity. Recently, the House passed a bill that would compensate landowners whose property has been devalued by more than 20% because of ESA restrictions and require the government to voluntarily purchase some of the land if there is a loss of more than half of the property value as a result of certain environmental laws (Healy, 1995). In addition, the bill would establish incentives to encourage property owners to protect habitat as well as shift much of the decision making to the states (Bornemeier, 1995). If passed in the Senate, the ESA will receive a major benefit: a wider range of support. Property owners will have more power and will offer less opposition.

Opponents of property rights argue, "Why should the landowners receive compensation or incentives for their land?" The answer is simply because they are currently alienated from the process. The common consensus among property owners can be summarized by Dick Christy, a Montana rancher: "I lost my rights as a private-property owner. . . . I'm a victim of the Endangered Species Act" (Chadwick, 1995, p. 14).

Instead of focusing on what kinds of breaks landowners might get if they follow ESA guidelines, the act concentrates on restrictions and penalties. According to Lacy and Lepper (1996a), it forbids the importing or exporting of listed species, including harassing, harming, pursuing, hunting, shooting, wounding, trapping, killing, capturing, or collecting of listed species, or possessing, selling, delivering, carrying, transporting, or shipping any endangered species unlawfully taken in the United States and its neighboring seas. The restrictions encompass a wide variety of activity, and penalties reflect this. Any violation of the ESA's restrictions can be met with fines totalling up to $100,000 and one year's imprisonment. Organizations can be fined up to $200,000. With so many rules, regulations, and such harsh penalties, no wonder landowners are frightened of the ESA.

It is apparent that the act must change if it is to maintain enough support to survive. To avoid legislative extinction, the ESA must be altered to include humans as well as endangered species. Blanket protection for all animals must cease; compensation must become as much a part of the Endangered Species Act as the "takings" clause. If David Trotter's land was protected by a reformed ESA, he wouldn't be "donating" it to the government without compensation. Perhaps his property never would have been taken in the first place, protected instead for economic reasons. He certainly would have

Carolyn ends her essay by returning to the beginning (a neat, if familiar, trick) trotting out Trotter once last time.

an overall better opinion of the Endangered Species Act. He might even become one of its supporters.

References

Bjerklie, D. (1992, May-June). The Endangered Species Act and its discontents. Technology Review, pp. 19-20.

Bronemeier, J. (1995, September 8). Bipartisan bid to revamp Endangered Species Act introduced in House. Los Angeles Times, p. A3.

Carpenter, B. (1995, July 10). Is he worth saving? U.S. News and World Report, pp. 43-45.

Chadwick, D. H. (1995, March). Dead or alive: The Endangered Species Act. National Geographic, pp. 2-41.

Counter-movement backs wise use. (1993, January). Christian Science Monitor, p. 11.

Fayhee, J. M. (1990, October). A hard act to follow: A primer to a most confusing piece of legislation. Backpacker, pp. 86-89.

Healy, M. (1995, March 4). House approves bill to give landowners relief. Los Angeles Times, p. A1.

Horton, T. (1992, March-April). The Endangered Species Act: Too tough, too weak, or too late? Audubon, pp. 68-74.

Endangered Species 16

Jost, K. (1995, June 16). Congress debates Endangered Species Act. Congressional Research Quarterly, pp. 515-519.

Lacy, K., & Lepper, C. (1996a). Brief history of the Endangered Species Act. Endangered Species [Online], 3 pgs. Available HTTP: http://gladstone.uorgeon.edu/~cait/abstract.htm [1996, May 15]

Lacy, K., & Lepper, C. (1996b). Case studies. Endangered Species [Online], 2 pages. Available HTTP: http://gladstone.uoregon.edu/~cait/casestudies.htm [1996, May 15]

Reiger, G. (1995, March). Law of nature. Field and Stream, pp. 16-17.

Lowercase letters are used to distinguish these two Internet documents, which are from the same authors and year.

APPENDIX C

□ ■ □

Tips for Researching and Writing Papers on Literary Topics

Before I turned to English teaching as a profession, I had a background in science. I had written lots of research papers on science-related topics—lobsters, oak tree hybridization, environmental education—but felt totally unprepared to write a research paper on a book or a poem. What do you write *about?* I wondered. I paced back and forth all night before my first literature paper was due.

I know now—for a paper on any topic—not to waste time, staring off into space and waiting for inspiration, but to pick up my pen and simply start writing. I trust that I'll discover what I have to say. I also know that a paper on a literary topic isn't really so different from those papers I wrote for other classes. All good papers simply involve taking a close look at something, whether it's the mating habits of a fly or an essay by George Orwell.

MINE THE PRIMARY SOURCE

What distinguishes a paper on a literary topic from others is where most of information—the details—come from. When I wrote the book on lobsters, that information came largely from interviews and a variety of published sources. When I wrote about the manhood issues raised by the characters in two Wallace Stegner novels, most of that information came *from the novels*.

A research paper about a story, poem, essay, or novel will usually use that work more than any other source. Literary papers rely heavily on *primary sources*. In addition to the literary works, primary sources might also include letters or interviews by the author related to those works.

In "Breaking the 'Utter Silence,'" the remarkable personal-response essay that comes later in this appendix, Kazuko Kuramoto writes exclusively about a single primary source: an essay by George Orwell. But she also manages to weave in another text, as well: her own experiences in Manchuria during World War II. The danger in writing such autobiographical/critical responses is that the writer can easily leave the written text behind completely and concentrate on her own story. But Kazuko doesn't do that. She returns again and again to particular phrases and ideas in Orwell's essay, and then she examines how they illuminate her own experiences. If there's one general weakness in student papers on literature, it's that they don't mine the works enough. The writer should not stray too far from what's in the poem, story, novel, or essay she is writing about.

This emphasis on mining primary sources means that the research strategy for a literary paper is often a little different than that for other topics. First, remember that your most important reading will not be what you dig up in the library. It will be your reading of the work you're writing about. Be an activist reader. Mark up the book (unless it's a library copy, obviously) or the story, underlining passages that strike you in some way, perhaps because they seem to reach below the surface and hint at what you think the writer is trying to say. Use a journal. The double-entry notetaking method described in "The Third Week" (Chapter 3) is a great way to explore your reactions to your reading. Use fastwriting as a way to find out what you think after each reading of the work; explore your reaction, rather than staring off into space, trying to figure out what you want to write about.

How do you know what you think until you see what you say?

SEARCH STRATEGIES

Though your close reading of the work you're writing about should be at the heart of your research, you can do other research, too. Most literary topics can be seen from several other basic angles. You can look at the *author* or what *critics* say about the author and his work. You may also discover that the author or work you're writing about fits into other recognized *categories* or *traditions*. In the most general sense, the work might be classified as British literature or American literature, but it also might fit into a subclass, such as African-American or feminist literature, or align with a particular regional school, like southern writers. Each of these classifications is a subject by itself and will be included in reference sources.

Let's look at a few key library sources for a paper on a literary topic.

Researching the Author

Biographies

Frequently, research on a literary topic begins with exploration of the author. Biographical sources on authors abound. Here are a few key reference works:

Authors' Biographies Index. Detroit: Gale, 1984–present. *A key source to 300,000 writers of every period.*

Biography Index: A Cumulative Index to Biographical Material in Books and Magazines. New York: Wilson, 1946–present. *Remarkably extensive coverage. Includes biographies, as well as autobiographies, articles, letters, obituaries, and the like.*

Contemporary Authors. Detroit: Gale, 1962–present. *Up-to-date information on authors from around the world; especially useful for obscure authors.*

Other helpful sources by the Wilson company, the familiar publishers of the *Reader's Guide,* include the following: *American Authors, 1600–1900; British Authors before 1800; British Authors of the Nineteenth Century;* and *European Authors, 1000–1900.*

Primary Bibliographies

It might be helpful to read additional works by the author you're researching. What else has he written? Biographies may tell you about other works, but they are often incomplete. For complete information, consult what's called a *primary bibliography,* or a bibliography *by* the author:

Bibliographic Index. New York: Wilson, 1937–present. *The "mother of all bibliographic indexes" lists works that have been published by and about the author.*

Bibliography of Bibliographies in American Literature. New York: Wilson, 1970. *Works by and about American authors.*

Index to British Literary Bibliography. Oxford: Clarendon, 1969–present. *Works by and about British authors.*

American Fiction: A Contribution Toward a Bibliography. San Marino: Huntington Library, 1957–present. *Entries on 11,000 novels, stories, and so on, indexed by author.*

Researching the Critics

What do other people say about your author and the work you're writing about? A thorough look at criticism and reviews is an important step in most research papers on literary topics, especially after you've begun to get a sense of what *you* think. Support from critics can be important evidence to bolster your own claims, or it can further your own thinking in new ways.

Several useful reference sources have already been mentioned. For example, so-called *secondary bibliographies,* or bibliographies *about* individual authors, are listed in the *Bibliographic Index* mentioned earlier. Check that. The most important index to check for articles about your author or her work is the *MLA International Bibliography*, mentioned in "The Third Week" (Chapter 3). This source is commonly available on CD-ROM. Other helpful references include:

Contemporary Literary Criticism. Detroit: Gale, 1973–present. *Excerpts of criticism and reviews published in the last twenty-five years.*

Magill's Bibliography of Literary Criticism. Englewood Cliffs, NJ: Salem, 1979. *Citations, not excerpts, of criticism of some 2,500 works.*

Book Review Index. Detroit: Gale, 1965–present. *Citations for tens of thousands of book reviews, including many obscure works.*

Current Book Review Citations. New York: Wilson, 1976–present. *Citations for reviews in about 1,200 publications.*

New York Times Book Review Index, 1896–1970. New York: Arno, 1973. *Great for a more historical perspective on literary trends; features about 800,000 entries.*

Researching the Genre or Tradition

What type of work, or *genre,* are you researching? A novel? A poem? Might it fit into some recognized category or tradition?

Another angle on your topic is to place your work or author in the context of similar works and authors. One place to begin is with general survey books, such as *The Oxford Companion to American or English Literature*. Other references are surveys of period literature, such as *English Literature in the Sixteenth Century*, as well as world literature, such as *History of Spanish American Literature* and *World Literature Since 1945*.

Within each of these broad literary landscapes are some smaller ones, each with its own reference sources. For example, within the broad topic American literature, there's a growing list of references to African-American literature. For example, *Afro-American Literature: The Reconstruction of Instruction* is a reference filled with essays on the place of African-American literature in literary history. Similarly, *American Indian: Language and Literature* lists 3,600 books and articles on American Indian literature and language.

Balay's *Guide to Reference Books* provides a helpful listing of sources that can help you place your topic in a larger context. But the best reference for research on a literary topic is this:

> Harner, James L. *Literary Research Guide.* 2nd ed. New York: MLA, 1993.

This amazing source reviews bibliographies, histories, indexes, surveys, and periodicals on every class and subclass of literature imaginable, from world literature to Chicano fiction. Buy the *Literary Research Guide* if you have to write a lot of papers on literature.

SAMPLE ESSAY: PERSONAL RESPONSE

Reading imaginative literature can be a deeply personal experience. It should be. As you're reading a novel, poem, or essay, pay attention to what moves you. A character or idea that gets your attention is often the launching place for a good paper. Your emotional response is your way into the author's work.

The personal-response essay, sometimes called *personal criticism* or *autobiographical criticism,* is simply one of many ways of writing about literature. But it's often one of the most satisfying ways to explore what you think about what you've read, and it's certainly an approach that's consistent with the emphasis in this book on trying to merge the *personal* with the *scholarly.*

That's why I've chosen Kazuko Kuramoto's personal response as an inspiring model of what can be done when the writer tries to think autobiographically about what she's read. If you try this, I think you'll find it difficult to do well. You might also discover, as I suggest later, that the personal-response essay can be a draft for a more fully researched critical essay. (That might be what your instructor has in mind.) Even so, in their own right, autobiographical/critical essays can be informative and insightful responses to literature, as I think you'll see.

"I Can Relate To It" Is Only a Start

Kazuko Kuramoto was an older woman in my Nonfiction Writing class. She was quiet, dignified, and quite lacking in confidence about her English language skills. Yet as you can see from the reading-response essay that follows, Kazuko writes elegantly about how a George Orwell essay on colonial Morocco in the 1930s challenges her to reexamine her own life as a girl in Imperial Japan during World War II. What is particularly effective about this essay is how Kazuko's autobiographical reflections are tied tightly to Orwell's text. She frequently uses phrases, quotations, and ideas from his essay as a means for understanding both her own experience *and* Orwell's. Kazuko is reading herself and Orwell's "Marrakech" simultaneously, and each illuminates the other.

The personal approach to writing about literature, then, is more than simply establishing that you can really relate to it, though that may be a start. As Kazuko demonstrates, personal criticism establishes an ongoing *conversation* with the text, interrogating it for ideas that shed light on personal experience, which then reflects back on the story or poem or essay, deepening or altering its meaning. This dialogue is not general or abstract; it is grounded in the specifics of the text and the specifics of the writer's experience.

Using the double-entry journal is one way to help get this conversation going (see Chapter 3). Another is *to read like a writer.* That means being a careful observer of your experience reading a literary text, especially those encounters with particular passages, scenes, characters, details, or ideas that make you wonder about yourself. Pay attention to things that lead you to the question, What does this say about me? In Kazuko's essay on "Marrakech," she discovers an answer to that question: Like Orwell, she once participated in the "utter silence" that is a precondition to imperialism.

"Breaking the 'Utter Silence'"* could be a draft for a more fully researched essay. Kazuko uses the primary text exclusively as a source in her response essay, though she also integrates the text of her own experience. Her essay could be further enriched in revision by using some of the biographical, bibliographic, and critical references already mentioned in this appendix. Where might further research be useful in this essay? What kinds of reference sources might Kazuko try?

(Note that Kazuko's paper follows MLA style—see Appendix A for more information.)

*"Breaking the 'Utter Silence': A Response to Orwell's 'Marrakech'" is reprinted with permission of Kazuko Kuramoto.

Kazuko Kuramoto

Professor Ballenger

English 201

29 February 1996

Breaking the "Utter Silence":

A Response to Orwell's "Marrakech"

In George Orwell's essay "Marrakech," which explores his experience as a British official in colonial Morocco, he does not argue his point with you. Orwell appeals to all your senses instead. As the essay opens, he takes you through the stench of a corpse that attracted the "cloud" (46) of flies from the restaurant. You follow a shabby funeral procession, breathing the smoldering air of the marketplace under the Moroccan sun--people and animals stained with sweat, souring fruits, and the dust of the traffic. You hear the wailing of short chant "over and over again" (46). You see the mounds of a Moroccan graveyard with "no gravestone, no identifying mark of any kind, . . . like a derelict building-lot." Then you see "how easily they die," and how "they rise out of the earth, . . . sweat and starve for a few years, and . . . sink back into the nameless mounds of the graveyard" (46).

Kazuko writes like Orwell: with rich detail. In this lead paragraph, the details come from Orwell's essay.

Then you understand and accept Orwell's point quite readily that "all colonial empires

Kuramoto 2

are in reality founded upon that fact. . . . Are they really the same flesh as you are? Do they even have names? Or are they merely a kind of undifferentiated brown stuff, as individual as bees or coral insects" (Orwell, "Marrakech" 46)? George Orwell does not argue his point. He paints it. When the curtain falls on this essay, you are left with dazzling impressions.

Here, Kazuko hints at her purpose: to face her own life as honestly as Orwell did in his essay.

Of course, I admire George Orwell for his skillful writing style: clear, concrete, and without frills and mannerisms. I dream of writing like George Orwell someday. But more so, I admire his honesty, his passion for truth and his "power of facing unpleasant facts" (qtd. in Smart 34).

However painful or ugly it may be, Orwell does not shy away from the truth. In "Marrakech," he first describes an Arab laborer, watching him feed a gazelle: "Finally [the laborer] said shyly in French: 'I could eat some of that bread.' I tore off a piece and he stowed it gratefully in some secret place under his rags. This man is an employee of the Municipality" (46). And then the Jews, who live in the Jewish quarters that reminds one of medieval ghettoes; windowless houses, sore-eyed children "in unbelievable numbers,

Kuramoto 3

like clouds of flies," and the narrow street where "there is generally running a little river of urine" (47). Then he points out the irony of the belief that the Jews are "the real rulers of this country. . . . They've got all the money. They control the banks, finance--everything" (47). He compares this to the burning of old women for witchcraft "when they could not even work enough magic to get themselves a square meal" (48). Indeed, he does not hide behind "the utter silence that is imposed on every Englishman in the East" (Orwell, "Shooting" 35). He speaks up. He shows the world the reality of imperialism.

Parenthetical citations should show clearly what page(s) quoted materials come from.

Here, Kazuko shifts to autobiography. Does this seem awkward, or does she keep her narrative anchored to Orwell's essay?

I recognize the "utter silence." I am a product of imperialism, a Japanese, born and raised in Dairen, Manchuria, when the area was one of Imperial Japan's colonies. As the third generation in Dairen, I was born into the society of Japanese supremacy and grew up believing in Japan's "divine mission" to "save" Asia from the evil hands of Western imperialists: British in India and Hong Kong, French in Indo-China, and Dutch in East Indies. Reading Orwell's essay "Marrakech" brought back many memories that I had long discarded, had preferred not to remember.

Kuramoto 4

The "utter silence" imposed on all Japanese in Japanese colonies is one of them. I did not recognize it then, but I do now from remembering my father. My memory of him is partly how he represented the generation of Japanese who have kept their "utter silence" to their graves. One particular incident depicts the "utter silence" clearly in my memory. It happened one day in the spring of 1945 in a small town at the border of the Japanese colony, where my father was the head of the Japanese government. My father and I visited the town's Shinto shrine to dedicate one minute of prayer for the war dead and also to pray for Japan's victory, as required by government. My father was a loyal Japanese. On our way back, we saw a group of Manchurian high school students. They were marching in an orderly military column, as required of all students at the time, Japanese or Manchurians. As they came closer to us, the teacher, who was leading the column, recognized my father. Suddenly the teacher ordered his troop "Attention!" followed by "Eye-e Right!" all in clear and loud Japanese. My father was astounded, to say the least. He quickly looked around to see if this formal military group salute was meant for him or someone nearby but saw no one. He

Kuramoto 5

straightened himself up and returned the salute, imitating military fashion as best as he could manage. I remained standing by him, dumbfounded.

"Wow, what a surprise . . ." I said, catching up to his suddenly quickened steps. I knew he was terribly embarrassed.

Notice how dialogue speeds things up.

"It's this uniform," he said somewhat curtly. He was wearing one of those government-ordered khaki "citizen's clothes" and the matching cap, closely resembling Japanese military uniform.

"Did that teacher take you for someone else, I wonder?" I said.

"No, he knows me."

"Oh, well, you are one of the highest-ranking people in town."

"I am only a civil servant," he cut me off short, almost angry.

"Yeah, but . . ." I swallowed the rest of the sentence--but we are Japanese . . .

Did I mean that Manchurians should salute all Japanese government officials just because they were Japanese? I now wonder, but I must have. The Japanese government in the Manchurian colony was the frontier symbol of Japan's international power, the power of "the Rising Sun." Why not? Salute to us. Salute to

Kuramoto 6

us all! Yet, on the other hand, I knew that the Japanese supremacy that my generation of Japanese in Dairen took for granted had always made my father uneasy. He had a reputation for being fair and considerate to his Manchurian subordinates and friends. He was well liked and respected among them, Japanese and Manchurians. He went out of his way to teach us children to treat local Chinese, the Manchurians, with respect, while he did not openly deny what we were taught at school: that Japanese were the almighty leaders of Asia.

We were the rulers, superior to all others--I had innocently believed it and had taken it for granted, while my father kept his "utter silence." Were we protected by the "utter silence" of the adults around us? Or were we deceived? Were the Manchurians simply invisible to the Japanese, as Orwell suggests was the case with the Moroccans to the Europeans?

It took Orwell several weeks before he noticed old women underneath the pile of firewood passing by, while he admits that "[he] had not been five minutes on Moroccan soil before [he] noticed the overloading of the donkeys and was infuriated by it"

Kuramoto 7

("Marrakech" 49). Does he mean to say that had the old women been white, or better yet British, he would have been infuriated? Orwell seems to blame the invisibility of the Moroccans on the color of their skin: "In a tropical landscape one's eye takes in everything except the human beings. . . . [I]t always misses the peasant hoeing at his patch. He is the same colour as the earth, and a great deal less interesting to look at." And again, ". . . But where the human beings have brown skins their poverty is simply not noticed" (48).

But this does not apply to the Japanese in colonial Dairen. Japanese and Chinese share the same skin color. Yet what Orwell says next is true of how it was with the Japanese in Dairen: "One could probably live here for years without noticing that for nine-tenths of the people the reality of life is an endless, back-breaking struggle to wring a little food out of an eroded soil" ("Marrakech" 48). We hardly noticed the Chinese. We were taught to be "nice and kind" to the native Chinese, the Manchurians. Yet under the strict segregation, we never had Chinese neighbors or Chinese classmates. The native Chinese were our domestic servants, coolies, and peddlers, who

Kazuko both complicates and extends the idea of invisibility in Orwell's essay.

lived in the areas where ordinary Japanese did not even think of going. They were invisible to us. How did this happen among people with the same skin color?

When a country surrenders to another, its people become one of the winner's possessions, along with the land and buildings, and they lose their individual identity. They live in the shadow of the ruthless oppressors. Invisible. And the rule of the mask, "He wears a mask, and his face grows to fit it" (Orwell, "Shooting" 38), applies equally to both the oppressor and the oppressed. They conspire with one another to make imperialism possible. The oppressed assume the role of the helpless and silent subjects, feeding the already dangerous hubris of the oppressor. And then, the "feeling of reverence before a white skin" (Orwell, "Marrakech" 50) existed among the colored, brown, yellow, or black, prostrating themselves before the Western power.

The white skin had represented the advancement of civilization for a long time, long enough to establish a superiority complex among the whites, an inferiority complex among the colored. Japan resisted it and became imperialist herself. She plunged into the

Kuramoto 9

world of "dog eats dog" when she awoke from the two hundred years of self-imposed national isolation. World War II in the Pacific stemmed from the fight among the imperialists: Western imperialists in the East against one small but refractory Eastern imperialist, Japan.

Orwell's "Marrakech" prodded me to think through my long-pending question: What is the imperialism that has toppled my life? I was possessed by Orwell's passion for truth, his power of "facing unpleasant facts," and read the essay with very personal interest. And now, perhaps, I not only recognize the "utter silence" of Orwell's imperialism but have found the courage to speak.

Kuramoto 10

Works Cited

Smart, William. Eight Modern Essayists. 6th
ed. New York: St. Martin's, 1995.

"George Orwell." Smart 31-34.

Orwell, George. "Marrakech." Smart 45-50.

---."Shooting an Elephant." Smart 35-41.

To simplify the "Works Cited" when documenting several selections from an anthology or reader, first list the complete citation for the book; then list the individual selections you've drawn from it. For those selections, list the author, title, name of the editor(s) of the reader, and page numbers.

SAMPLE RESEARCH ESSAY ON A LITERARY TOPIC

In the sample paper that follows—"Metamorphosis, the Exorcist, and Oedipus"*—Karoline chooses a less personal approach but one that is still effective. Through careful use of the primary text—Kafka's short story "The Metamorphosis"—along with just two other sources—a biography and a collection of Kafka's diary entries—Karoline shows how the author and his story merge. She remains behind the scenes in this paper—at least, much more so than Kazuko did in her response essay on Orwell. But Karoline's paper still relays a sense of a purposeful writer, shaping and shaving the material, someone who has a clear sense of what she wants to say and who wants us to understand it.

(Karoline's paper also uses MLA conventions.)

*"Metamorphosis, the Exorcist, and Oedipus" is reprinted with permission of Karoline A. Fox.

Fox 1

Karoline Ann Fox

Professor Dethier

English 401

15 December 1991

Metamorphosis, the Exorcist,
and Oedipus

According to Ernst Pawel, Franz Kafka's "The Metamorphosis" goes beyond "standard categories of literary criticism; it is a poisoned fairy tale about the magic of hats and the power of hypocrisy . . . charting the transmogrification of a lost soul in a dead bug" (279). Kafka's tale is more than a literary work. It is a frighteningly realistic representation of the most desperate and obsessing fears of its author. Disguised behind the all-too literal shell of the character Gregor Samsa, Kafka attempts to exorcise the demons of his haunted childhood.

Karoline's thesis is clearly stated in the lead. In a short essay such as this, it's helpful to state the thesis early.

In one of his many diary entries, Kafka described writing as "the revealing of oneself to excess; that utmost self-revelation and surrender . . ." (Memory 72), yet he admits that this surrender fulfills "a great yearning to write all my anxiety entirely out of me, write into the depths of the paper just as it

When citing 2 or more sources by the same author, abbreviate the titles. Note that Kafka's name is mentioned in text, so it's not included in the parenthetical citation.

Fox 2

comes out of the depths of me . . ." (38). In "The Metamorphosis," Kafka clearly makes no attempt to hide his deepest anxieties but reveals them with deliberate vindictiveness.

Karoline effectively weaves passages from different parts of the story into this paragraph.

There is, for example, a striking resemblance between Kafka's father and Gregor's father in the story. Like Herrmann Kafka, Mr. Samsa seems a tyrannical giant, whose violence overwhelms any sympathy for his unfortunate son. In fact, Mr. Samsa's first reaction to his son's condition is to knot "his fist with a fierce expression on his face as if he meant to knock Gregor back into his room . . ." (869). And even Gregor's kindest intentions are misunderstood by his father, who "pitilessly . . . drove him back, hissing and crying 'Shoo!' like a savage," and ultimately sent him flying into his room, "bleeding freely" (871).

Cite only page numbers when it's clear from the text where borrowed material has come from.

Though Herrmann never abused the young Kafka, his son had an unnatural fear of the man. Remembering a time when his father punished Kafka by locking him out on a balcony, his son later wrote, "Even years afterward I suffered from the tormenting fancy that the huge man, my father, the ultimate authority, would come almost for no reason at all and take me out of bed in the

Fox 3

night and carry me onto the [balcony], and that meant I was mere nothing for him" (Memory 10).

Kafka's view of his father as a rival for most, if not all, of his life indicates an oedipal complex, according to Pawel (15). Kafka's mother, Julie, was completely devoted to her husband but neglected her son, who was taken care of by servants until old enough to care for himself. Kafka did not take out his anger for this neglect on his absent mother but on the recipient of her affection, Herrman.

Gregor, in "The Metamorphosis," shares this jealousy. He witnesses an almost erotic reunion between his parents as he sees "his mother rushing toward his father, leaving one after another behind her on the floor her loosened petticoats, stumbling over her petticoats to his father and embracing him in complete union . . ." (881-882). Yet his mother seems saintly, despite her loosened petticoats, "with her hands clasped around his father's neck . . . she begged for her son's life" (882). Gregor's mother is his angel of mercy, and only she harbors hope that he will one day become human again. She refuses to remove the furniture from his room, saying

Fox 4

". . . when he comes back to us he will find everything unchanged and be able all the more easily to forget what has happened in between" (878).

Despite Kafka's idealized vision of his mother, the author and his mother were never very close. Julie Kafka was always at Herrmann's beck-and-call. As a boy of five, Franz saw the birth of first one and then another sibling as a threat to his monopoly of what was almost nonexistent maternal attention. With the successive deaths of his sibling rivals, Kafka suffered from traumatic guilt, which he fought the rest of his life (Pawel 63).

Throughout, Karoline uses biographical information about the author to reflect against material in his short story, each illuminating the other.

He felt no such rivalry when three more sisters were born, years later. Instead, Kafka was happy to have the companionship and became particularly fond of his youngest sister, Ottilie. In "The Metamorphosis," Gregor also has a younger sister for whom he is willing to sacrifice a portion of his salary so she can study violin at the Conservatorium. It is Grete, who "in the goodness of her heart" (873) brings Gregor food and cleans his room.

Grete's compassion fades, however, because the young girl can't endure her brother's appearance, making Gregor "realize how repulsive the sight of him still was to her, and that it was bound to go on being repulsive" (876). Grete later abandons her good intentions toward her brother but keeps up her care, enjoying a kind of selfish martyrdom.

Later, even that disappears and Grete's rejection of her brother is complete. When Gregor hears Grete play the violin, he remembers his longing to send her to the Conservatorium and, weak from pain and near starvation, crawls toward the sound of the music "determined to push forward till he reached his sister, to pull at her skirt and so let her know that she was to come into his room with her violin, for no one . . . appreciated her playing as he would appreciate it" (887). Instead of seeing the affection, Grete is horrified, saying ". . . we must try to get rid of it" (888).

Like Gregor, Kafka felt himself trapped within a hatefully repulsive shell, writing "it is certain that a major obstacle to my progress is my physical condition. Nothing can be accomplished with such a body" (Memory 37). He feels incapable of being loved by anyone. Desperately, he tried to conceal his need for affection with indifference, claiming "the sorrows and joys of my relatives bore me to the very soul" (89) and "all parents want to do is drag one down to them, back to the old days from which one longs to fill oneself and escape" (60). This denial of his need for love led Kafka to severe depression, suicidal self-hatred, insecurity, self-pity, hypochondria, and serious physical ailments (Pawel 191).

Even as a young man, Kafka put himself down, insisting that he would fail in every endeavor: "He sincerely believed himself to be incompetent, lazy, forgetful, clumsy, badly dressed, incoherent" (Pawel 53). Admitting "success . . . did not inspire confidence; on the contrary, I was always convinced . . . that the more I accomplished, the worse off I would be in the end" (Kafka, <u>Memory</u> 26).

This incessant self-doubt, however, served as a safety net for Kafka. If he insisted he would fail, he couldn't be held accountable when he did. If he succeeded, then he could enjoy his success without setting unreasonable goals for himself. Kafka's supersensitive ego could never have withstood the defeat of both failure and wounded pride because he so desperately sought approval of his friends and his enemies alike.

Fox 7

Writing becomes the glue that holds Kafka together. Again and again, he curses his incompetence at it, arguing "God does not want me to write while I, I have to write. And so there is a constant tug-of-war" (Memory 7). It was a tug-of-war in which he had to participate, because Kafka clung to his writing more than he did to life itself: "The novel is me, my stories are me. . . . It is through my writing that I keep a hold of life" (Memory viii).

Not surprisingly, Kafka's family did not approve of his writing. When he wrote his father about his plans to be a writer, it's likely that the young author found some pleasure in his father's disappointment. But his father never read it. His mother did not give it to him, thinking that her son would outgrow his interest in writing (Memory ix). Julie Kafka's lack of understanding left her son embittered, and he later wrote in his diary a dialogue between himself and his mother which ended, "Certainly, you are all strangers to me, it is only blood that connects us, but that never shows itself" (Memory 91).

Kafka wrote that "what we need are books that affect us like some really grievous

Fox 8

misfortune, like the death of one whom we loved more than ourselves, as if we were banished to distant forests, away from everybody, like a suicide" (Memory 7). Of "The Metamorphosis" he said, "[T]he more I write, the more I liberate myself" (Memory 61). This liberation came from his ability to put his life into his story and exorcise the demons of his childhood. Paradoxically, by doing so, he was able to hold onto that life.

"The Metamorphosis" embodies those conflicting sentiments. In it, we see more than the dried and withered empty shell of a life. We see a man, Kafka, lying in the darkest corner of a filthy room, unable to move into the light and accept the love he so desperately needs in order to remain alive.

Karoline restates her thesis but in a fresh way.

Fox 9

Works Cited

Karoline uses just 3 sources, but vigorous use of the primary source—the Kafka short story—throughout the paper helps make it meaty.

Use three dashes to show another work by the author in the preceding citation.

Kafka, Franz. I Am a Memory Come Alive. Ed. Nahum N. Glatzer. New York: Schocken Books, 1974.

---. "The Metamorphosis." Fiction 100. Ed. James Pickering. New York: Macmillan, 1988.

Pawel, Ernst. The Nightmare of Reason. New York: Farrar, 1984.

Index

Note: Boldface numbers indicate pages with illustrations.